Compendium of Plant Genomes

Series Editor

Chittaranjan Kole, Raja Ramanna Fellow, Government of India,
ICAR-National Research Center on Plant Biotechnology, Pusa,
New Delhi, India

Whole-genome sequencing is at the cutting edge of life sciences in the new millennium. Since the first genome sequencing of the model plant Arabidopsis thaliana in 2000, whole genomes of about 100 plant species have been sequenced and genome sequences of several other plants are in the pipeline. Research publications on these genome initiatives are scattered on dedicated web sites and in journals with all too brief descriptions. The individual volumes elucidate the background history of the national and international genome initiatives; public and private partners involved; strategies and genomic resources and tools utilized; enumeration on the sequences and their assembly; repetitive sequences; gene annotation and genome duplication. In addition, synteny with other sequences, comparison of gene families and most importantly potential of the genome sequence information for gene pool characterization and genetic improvement of crop plants are described.

Interested in editing a volume on a crop or model plant?
Please contact Prof. C. Kole, Series Editor, at ckoleorg@gmail.com

More information about this series at http://www.springer.com/series/11805

Malali Gowda · Ambardar Sheetal ·
Chittaranjan Kole
Editors

The Neem Genome

Editors
Malali Gowda
TransDisciplinary University Yelahanka
Bengaluru, Karnataka, India

Ambardar Sheetal
TransDisciplinary University Yelahanka
Bengaluru, Karnataka, India

Chittaranjan Kole
Raja Ramanna Fellow,
Government of India
ICAR-National Research Center
on Plant Biotechnology, Pusa
New Delhi, India

ISSN 2199-4781 ISSN 2199-479X (electronic)
Compendium of Plant Genomes
ISBN 978-3-030-16121-7 ISBN 978-3-030-16122-4 (eBook)
https://doi.org/10.1007/978-3-030-16122-4

© Springer Nature Switzerland AG 2019
This work is subject to copyright. All rights are reserved by the Publisher, whether the whole or part of the material is concerned, specifically the rights of translation, reprinting, reuse of illustrations, recitation, broadcasting, reproduction on microfilms or in any other physical way, and transmission or information storage and retrieval, electronic adaptation, computer software, or by similar or dissimilar methodology now known or hereafter developed.
The use of general descriptive names, registered names, trademarks, service marks, etc. in this publication does not imply, even in the absence of a specific statement, that such names are exempt from the relevant protective laws and regulations and therefore free for general use.
The publisher, the authors and the editors are safe to assume that the advice and information in this book are believed to be true and accurate at the date of publication. Neither the publisher nor the authors or the editors give a warranty, express or implied, with respect to the material contained herein or for any errors or omissions that may have been made. The publisher remains neutral with regard to jurisdictional claims in published maps and institutional affiliations.

This Springer imprint is published by the registered company Springer Nature Switzerland AG
The registered company address is: Gewerbestrasse 11, 6330 Cham, Switzerland

This book series is dedicated to my wife Phullara, and our children Sourav, and Devleena
Chittaranjan Kole

Preface to the Series

Genome sequencing has emerged as the leading discipline in the plant sciences coinciding with the start of the new century. For much of the twentieth century, plant geneticists were only successful in delineating putative chromosomal location, function, and changes in genes indirectly through the use of a number of "markers" physically linked to them. These included visible or morphological, cytological, protein, and molecular or DNA markers. Among them, the first DNA marker, the RFLPs, introduced a revolutionary change in plant genetics and breeding in the mid-1980s, mainly because of their infinite number and thus potential to cover maximum chromosomal regions, phenotypic neutrality, absence of epistasis, and codominant nature. An array of other hybridization-based markers, PCR-based markers, and markers based on both facilitated construction of genetic linkage maps, mapping of genes controlling simply inherited traits, and even gene clusters (QTLs) controlling polygenic traits in a large number of model and crop plants. During this period, a number of new mapping populations beyond F2 were utilized and a number of computer programs were developed for map construction, mapping of genes, and mapping of polygenic clusters or QTLs. Molecular markers were also used in the studies of evolution and phylogenetic relationship, genetic diversity, DNA fingerprinting, and map-based cloning. Markers tightly linked to the genes were used in crop improvement employing the so-called marker-assisted selection. These strategies of molecular genetic mapping and molecular breeding made a spectacular impact during the last one and a half decades of the twentieth century. But still, they remained "indirect" approaches for elucidation and utilization of plant genomes since much of the chromosomes remained unknown and the complete chemical depiction of them was yet to be unraveled.

Physical mapping of genomes was the obvious consequence that facilitated the development of the "genomic resources" including BAC and YAC libraries to develop physical maps in some plant genomes. Subsequently, integrated genetic–physical maps were also developed in many plants. This led to the concept of structural genomics. Later on, emphasis was laid on EST and transcriptome analysis to decipher the function of the active gene sequences leading to another concept defined as functional genomics. The advent of techniques of bacteriophage gene and DNA sequencing in the 1970s was extended to facilitate sequencing of these genomic resources in the last decade of the twentieth century.

As expected, sequencing of chromosomal regions would have led to too much data to store, characterize, and utilize with the-then available computer software could handle. But the development of information technology made the life of biologists easier by leading to a swift and sweet marriage of biology and informatics, and a new subject was born—bioinformatics.

Thus, the evolution of the concepts, strategies, and tools of sequencing and bioinformatics reinforced the subject of genomics—structural and functional. Today, genome sequencing has traveled much beyond biology and involves biophysics, biochemistry, and bioinformatics!

Thanks to the efforts of both public and private agencies, genome sequencing strategies are evolving very fast, leading to cheaper, quicker, and automated techniques right from clone-by-clone and whole-genome shotgun approaches to a succession of second-generation sequencing methods. The development of software of different generations facilitated this genome sequencing. At the same time, newer concepts and strategies were emerging to handle sequencing of the complex genomes, particularly the polyploids.

It became a reality to chemically—and so directly—define plant genomes, popularly called whole-genome sequencing or simply genome sequencing.

The history of plant genome sequencing will always cite the sequencing of the genome of the model plant Arabidopsis thaliana in 2000 that was followed by sequencing the genome of the crop and model plant rice in 2002. Since then, the number of sequenced genomes of higher plants has been increasing exponentially, mainly due to the development of cheaper and quicker genomic techniques and, most importantly, the development of collaborative platforms such as national and international consortia involving partners from public and/or private agencies.

As I write this preface for the first volume of the new series "Compendium of Plant Genomes," a net search tells me that complete or nearly complete whole-genome sequencing of 45 crop plants, eight crop and model plants, eight model plants, 15 crop progenitors and relatives, and 3 basal plants is accomplished, the majority of which are in the public domain. This means that we nowadays know many of our model and crop plants chemically, i.e., directly, and we may depict them and utilize them precisely better than ever. Genome sequencing has covered all groups of crop plants. Hence, information on the precise depiction of plant genomes and the scope of their utilization are growing rapidly every day. However, the information is scattered in research articles and review papers in journals and dedicated Web pages of the consortia and databases. There is no compilation of plant genomes and the opportunity of using the information in sequence-assisted breeding or further genomic studies. This is the underlying rationale for starting this book series, with each volume dedicated to a particular plant.

Plant genome science has emerged as an important subject in academia, and the present compendium of plant genomes will be highly useful to both students and teaching faculties. Most importantly, research scientists involved in genomics research will have access to systematic deliberations on the plant genomes of their interest. Elucidation of plant genomes is of interest not only for the geneticists and breeders, but also for practitioners of an array of plant science disciplines, such as taxonomy, evolution, cytology, physiology, pathology, entomology, nematology, crop production, biochemistry, and obviously bioinformatics. It must be mentioned that information regarding each plant genome is ever-growing. The contents of the volumes of this compendium are, therefore, focusing on the basic aspects of the genomes and their utility. They include information on the academic and/or economic importance of the plants, description of their genomes from a molecular genetic and cytogenetic point of view, and the genomic resources developed. Detailed deliberations focus on the background history of the national and international genome initiatives, public and private partners involved, strategies and genomic resources and tools utilized, enumeration on the sequences and their assembly, repetitive sequences, gene annotation, and genome duplication. In addition, synteny with other sequences, comparison of gene families, and, most importantly, the potential of the genome sequence information for gene pool characterization through genotyping by sequencing (GBS) and genetic improvement of crop plants have been described. As expected, there is a lot of variation of these topics in the volumes based on the information available on the crop, model, or reference plants.

I must confess that as the series editor, it has been a daunting task for me to work on such a huge and broad knowledge base that spans so many diverse plant species. However, pioneering scientists with lifetime experience and expertise on the particular crops did excellent jobs editing the respective volumes. I myself have been a small science worker on plant genomes since the mid-1980s and that provided me the opportunity to personally know several stalwarts of plant genomics from all over the globe. Most, if not all, of the volume editors are my longtime friends and colleagues. It has been highly comfortable and enriching for me to work with them on this book series. To be honest, while working on this series I have been and will remain a student first, a science worker second, and a series editor last. And I must express my gratitude to the volume editors and the chapter authors for providing me the opportunity to work with them on this compendium.

I also wish to mention here my thanks and gratitude to the Springer staff particularly Dr. Christina Eckey and Dr. Jutta Lindenborn for the earlier set of volumes and presently Ing. Zuzana Bernhart and Dr. Anette Lindqvist for all their timely help and support.

I always had to set aside additional hours to edit books besides my professional and personal commitments—hours I could and should have given to my wife, Phullara, and our kids, Sourav and Devleena. I must mention that they not only allowed me the freedom to take away those hours from them but also offered their support in the editing job itself. I am really not sure whether my dedication of this compendium to them will suffice to do justice to their sacrifices for the interest of science and the science community.

New Delhi, India   Prof. Chittaranjan Kole

Preface to the Volume

Neem (*Azadirachta indica* A. Juss) is a member of the mahogany family, Meliaceae. Neem trees are attractive broad-leaved evergreens that can grow up to 30 m tall and 2.5 m in girth. A Neem tree normally begins bearing fruit after 3–5 years and becomes fully productive in 10 years and it may live for more than two centuries. The tree is easily propagated—both sexually and vegetatively. It can be planted using seeds, seedlings, saplings, root suckers, or tissue culture. The Neem tree was first discovered in India about 4500 years ago. Neem is native to India and Burma, and also found in Bangladesh, Pakistan and Nepal. The tree is said to grow "almost anywhere" in the lowland tropics. However, it generally performs best in areas with annual rainfalls of 400–1200 mm. It thrives under the hottest conditions, where maximum shade temperature may soar past 50 °C, but it will not withstand freezing or extended cold. The taproot (at least in young specimens) may be as much as twice the height of the tree.

The Neem research was started in the 1920s by the Indian scientists; Neem's ability to repel insects was reported in the scientific literature in 1926–1929 (National Research Council 1992). Neem ingredients are applied in Ayurveda, Unani, homeopathy, and modern medicine for the treatment of many infectious, metabolic, or cancer diseases (Mohammad 2016). Neem products have been used for centuries in the field of agriculture and medicine, and the plant extracts of Neem have enormous potential to influence modern agrochemical research. Neem has been well known for various medicinal properties like anti-inflammatory, antipyretic, antihistamine, antifungal, anti-bacterial, anti-ulcer, analgesic, antiarrhythmic, anti-tubercular, antimalarial, diuretic, spermicide, anti-arthritic, anti-protozoal, insect repellent, anti-feedant, and anti-hormonal properties. Neem has one of the richest sources of secondary metabolites in nature, and over 300 compounds have been isolated from different parts of the Neem plant.

In this context, this book compiles up-to-date information on research and development related to the Neem plant. This book comprises a total of 12 chapters, which starts with the traditional practices and knowledge of Neem (Chap. 1) and ends with the Neem microbiome (Chap. 12). The first two chapters (1 and 2) provide a general introduction to the Neem, traditional practices related to Neem as documented in Ayurveda and India's mythology. Chapter 3 enumerates various secondary metabolites produced by Neem plant, which are crucial for multifunctional properties like anti-oxidation, anti-inflammation, antimalarial, and anti-carcinogenic activities. Chapter 4 describes application of Neem and its products in agriculture. Chapter 5 provides the information about phylogeny of Neem and related species in the Meliaceae family based on genomic resources. Chapter 6 summarizes various strategies and tools for sequencing the genome and transcriptome of Neem. Chapter 7 provides a comprehensive knowledge on assembly of nuclear genome, organelle genome (chloroplast and mitochondria), and transcriptome of Neem plant. Chapters 8 and 9 provide information regarding the presence of the repetitive sequences in the Neem genome and annotation of genome. Chapter 10 elucidates on the synteny analysis and comparison of gene families with related plant species for better understanding of gene evolution in Neem. Chapter 11 gives the detailed information about the tissue culture techniques to produce a triploid Neem plant. The content of this book ends with Chap. 12 which describes various endophytes inhabiting in Neem plant and producing the similar secondary metabolites as Neem plant. Altogether, this book will serve as a primary resource material for researchers, breeders, and students working on Neem plant.

These chapters have been authored by 13 scientists from different research institutes and universities in India. We express our thanks to them for their contributions and cooperation for this book project. Experts in the field reviewed each chapter, and thus we are thankful for their efforts to improve the quality of this compilation. Dr. Malali Gowda and Dr. Ambardar Sheetal express their thanks and high gratitude to Prof. Chittaranjan Kole, Series Editor of the "Compendium of Plant Genomes," for giving the opportunity to co-edit this book and for his constant support and guidance right from the inception till publication of this book on *The Neem Genome*. The editors also acknowledge the help from all the staff of Springer Nature at all the stages.

Bengaluru, India	Prof. Malali Gowda
Bengaluru, India	Dr. Ambardar Sheetal
New Delhi, India	Prof. Chittaranjan Kole

References

National Research Council (1992) Neem: a tree for solving global problems. The National Academies Press, Washington, DC. https://doi.org/10.17226/1924

Mohammad AA (2016) Therapeutics role of Azadirachta indica (Neem) and their active constituents in diseases prevention and treatment. Evid-Based Complement Altern Med 2016, Article ID 7382506, 11 p. http://dx.doi.org/10.1155/2016/7382506

Contents

1 **Neem: Traditional Knowledge from Ayurveda** 1
S. N. Venugopalan Nair, Naveen Shilpa, Thomas Vargheese and I. F. Tabassum

2 **Heritage of Neem–Peepal Tree Resides a Profound Scientific Facts** 13
K. L. Ashalatha and Malali Gowda

3 **Method to Quantify Plant Secondary Metabolites: Quantification of Neem Metabolites from Leaf, Bark, and Seed Extracts as an Example** 21
Kannan Rangiah and Malali Gowda

4 **Utilization of Neem and Neem Products in Agriculture** 31
Rishu Sharma and Chittaranjan Kole

5 **Phylogeny of Neem and Related Species in the Meliaceae Family** 49
Nagesh A. Kuravadi and Malali Gowda

6 **Strategies and Tools for Next Generation Sequencing** 53
Nagesh A. Kuravadi and Malali Gowda

7 **Neem Genome Assembly** 59
Nagesh A. Kuravadi and Malali Gowda

8 **Repetitive Sequences** 65
Nagesh A. Kuravadi and Malali Gowda

9 **Neem Genome Annotation** 73
Nagesh A. Kuravadi and Malali Gowda

10 **Comparison of Gene Families and Synteny Analysis from Neem Genome** 93
Nagesh A. Kuravadi and Malali Gowda

11　**Neem Tissue Culture**............................... 99
　　Divya Mohan, Ashmita J. Tontanahal, B. N. Sathyanarayana
　　and Malali Gowda

12　**Neem Microbiome**................................. 111
　　Varalaxmi B. Agasimundin, Kannan Rangiah,
　　Ambardar Sheetal and Malali Gowda

Contributors

Varalaxmi B. Agasimundin Centre for Functional Genomics and Bioinformatics, TransDisciplinary University, Bengaluru, India;
Centre for Cellular and Molecular Platforms, Bengaluru, India

K. L. Ashalatha Centre for Functional Genomics and Bioinformatics, The University of TransDisciplinary Health Sciences and Technology, Bengaluru, India

Malali Gowda Center for Functional Genomics and Bio-Informatics, The University of TransDisciplinary Health Sciences and Technology, Foundation for Revitalization of Local Health Traditions, Bengaluru, India

Chittaranjan Kole National Research Centre on Plant Biotechnology, New Delhi, India

Nagesh A. Kuravadi Centre for Cellular and Molecular Platforms, National Centre for Biological Sciences, Bengaluru, Karnataka, India

Divya Mohan The University of TransDisciplinary and Health Sciences, Bengaluru, Karnataka, India;
Department of Horticulture, University of Agricultural Sciences, Bengaluru, Karnataka, India

Kannan Rangiah Centre for Cellular and Molecular Platforms, Bengaluru, India;
Food Safety and Analytical Quality Control Laboratory, CSIR-Central Food Technological Research Institute (CFTRI), Mysore, Karnataka, India

B. N. Sathyanarayana Department of Horticulture, University of Agricultural Sciences, Bengaluru, Karnataka, India

Rishu Sharma Department of Plant Pathology, BC Agricultural University, Mohanpur, West Bengal, India

Ambardar Sheetal Centre for Functional Genomics and Bioinformatics, TransDisciplinary University, Bengaluru, India;
Centre for Cellular and Molecular Platforms, Bengaluru, India

Naveen Shilpa The University of TransDisciplinary Health Sciences and Technology (TDU), The Center for Traditional Knowledge Informatics and Data Sciences, Yelahanka, Bengaluru, India

I. F. Tabassum The University of TransDisciplinary Health Sciences and Technology (TDU), The Center for Traditional Knowledge Informatics and Data Sciences, Yelahanka, Bengaluru, India

Ashmita J. Tontanahal Department of Horticulture, University of Agricultural Sciences, Bengaluru, Karnataka, India

Thomas Vargheese The University of TransDisciplinary Health Sciences and Technology (TDU), The Center for Traditional Knowledge Informatics and Data Sciences, Yelahanka, Bengaluru, India

S. N. Venugopalan Nair The University of TransDisciplinary Health Sciences and Technology (TDU), The Center for Traditional Knowledge Informatics and Data Sciences, Yelahanka, Bengaluru, India

Abbreviations

2,4-DNP	2,4-Dinitrophenol
Aza	Azadirachtin
Azadi	Azadiradione
BLAST	Basic local alignment search tool
bp	Base pair
CDS	Coding sequence
CGIs	CpG islands
CH	Casein hydrolysate
CID	Collision-induced dissociation
cp	Chloroplast
CTAB	Cetyltrimethylammonium bromide
DEG	Differentially expressed gene
dNTP	Deoxyribonucleotide triphosphate
dsDNA	Double-stranded DNA
E/H-azadi	Epoxy or hydroxy-azadiradione
EC	Enzyme Commission
EST	Expressed sequence tag
FACS	Fluorescence-activated cell sorting
FISH	Fluorescent in situ hybridization
GA	Gibberellic acid
GO	Gene ontology
GS	Genomic selection
IR	Inverted repeat
ITS	Internal transcribed sequence
KEGG	Kyoto Encyclopedia of Genes and Genome
KO	KEGG orthology
LA	Luria Bertani Agar
LINEs	Long interspersed nuclear elements
LTRs	Long terminal repeats
ME	Mate-pair
MEGA	Molecular evolutionary genetic analysis
MISA	MIcroSAtellite tool
N50	Minimum contig length
NCBI	National Center for Biotechnology Information
NF	Not found
NGS	Next-generation sequencing
nt	Nucleotide

Orf/ORF	Open reading frame
PAGE	Polyacrylamide gel electrophoresis
PCR	Polymerase chain reaction
PDA	Potato dextrose agar
PE	Pair end/ paired-end
QC	Quality Check
RbcL	Ribulose-1,5-bisphosphate carboxylase/oxygenase large subunit
RE	Repetitive element
RIN	RNA integrity value
ROS	Reactive oxygen species
RPKM	Reads per kilobase per million
rRNA	Ribosomal RNA
RuBisCo	Ribulose 1,5-bisphosphate carboxylase/oxygenase
SAM	Sequence alignment map
SINE	Short interspersed nuclear element
SNP	Single-nucleotide polymorphism
SRM	Selected reaction monitoring
SSR	Simple sequence repeat
STR	Short tandem repeat
TE	Transposable element
TF	Transcription factor
UHPLC-MS	Ultrahigh-performance liquid chromatography-mass spectrometry

Neem: Traditional Knowledge from Ayurveda

S. N. Venugopalan Nair, Naveen Shilpa, Thomas Vargheese and I. F. Tabassum

Abstract

Ancient Indian medical literature is said to be one of the world's most well-documented and incomparable evidences of the extreme wealth of Indian knowledge systems which dates back between 1500 BCE and 1900 CE. Detailed theoretical foundations and resources from Ayurveda, Unani, Siddha, and Tibetan medicine on medicinal plants, disease diagnosis, and treatment methods have long existed with considerable documented evidences. While on the other hand, undocumented knowledge is widely practised by hundreds of tribal communities across India and is often regarded as folk medical knowledge. Surpassing 6500 well-documented medicinal plants, India holds a treasure house of knowledge in this regard and Neem is one of the most important medicinal plant being widely used by all medicinal systems. This chapter aims to capture the essence of traditional knowledge on Neem, as documented in Ayurveda with detailed references. Various examples with detailed citations and descriptions, translated from ancient Acharyas such as *Charaka, Susrutha,* and *Vaghbata* have been covered. Vernacular names and Sanskrit synonyms of Neem along with its descriptive meaning illustrates the importance and popularity of Neem at a wider scale. The sheer depth of Indian Knowledge Systems is demonstrated with Sanskrit slokas and its approximate English meaning with the explanation on the use of various parts of the Neem as well as simple and compound medicinal formulations. Highly informative data sources and bibliography with supportive evidences authentic in nature have also been provided for further referencing.

1.1 Introduction

Indian medical knowledge systems including Ayurveda, Siddha, and Unani have extensively documented neem-related traditional knowledge. The plant is also widely used in folk medicine.

Caraka (The author of Caraka Samhita, believed to have been compiled between 1500 BCE and 200 CE) explains in Sanskrit "*yōgamāsām tu yō vidyāt dēśa kālōpa pāditam, puruṣam puruṣam vīkṣya sañjēyō sa bhiṣaguttama:*" (Trikamji 1992) that "He is the best of physicians who knows how to administer the medicine in accordance with their region

S. N. Venugopalan Nair (✉) · N. Shilpa ·
T. Vargheese · I. F. Tabassum
The University of TransDisciplinary Health Sciences and Technology (TDU), The Center for Traditional Knowledge Informatics & Data Sciences, 74/2, Jarakabande Kaval, Attur P.O., Yelahanka, Bengaluru 560064, India
e-mail: venu.gopal@tdu.edu.in
URL: http://www.tdu.edu.in

© Springer Nature Switzerland AG 2019
M. Gowda et al. (eds.), *The Neem Genome*, Compendium of Plant Genomes,
https://doi.org/10.1007/978-3-030-16122-4_1

(habitation and procurement of medicinal plants) and time and *prakṛti* (Psycho-somatic constitution of an individual, i.e., phenotypical or genetic typing) of each person individually" (Nair 2015).

The plant *Nimba* or Neem has been reported by the earliest documents of Ayurveda by Caraka, Susruta (1500 BCE–400 CE), and later by Vaghbata (500 CE) and then followed by many authors of medical lexicons (*nighantus*) down the centuries. The bibliographic database of the Trans-Disciplinary University (TDU) shows that the word *Nimba* itself has 448 references across 20 major classical texts of Ayurveda (INMEDPLAN 2018), written during the period 1500 BCE–1900 CE. This indicates the extent of information available in the Ayurvedic literature. The Sanskrit word "*nimba*" means (*Nimbati sincati svasthyamiti, svastha vrtikaramiti yavat*) that which gives health (Sharma 2007). It also denotes the ability of the drug to provide health and hygiene in daily life.

References to *Nimba* are found not only in medical texts but are also seen in the Vedic literature and *puranas* (epics). In *Ramayana*, *Nimba* has been mentioned as one of the main 10 trees of Nandavana. The period of Ramayana is *Tretayuga*, which is fixed around 3000 BCE. Nimba has been well cited in the popular scriptures related to various disciplines like *brhat samhita*, and *padma purana* and many other Sanskrit literature works and poetry including *kadambari*. In Indian astrology, worship of neem tree is indicated for people belong to the star *purvabhadra* to ward away ill effects.

It is interesting to note that the neem tree has more than 50 Sanskrit synonyms in Ayurvedic literature (INMEDPLAN 2018). The polynomial nomenclature system followed by the *acharayas* (authors of ancient Indian literature) has the purpose to define a drug in terms of its various morphological and clinical features. Neem is called *arista* because it eradicates diseases. As crows eat fruits of neem, it is called *kakaphala*. It induces vomiting (*chardana*) and also used as antiemetic (*chardighna*). it is termed as "*hingu niryasa*" because the tree exudes a gum similar to *hingu* (asafetida). The anti-inflammatory property of the neem is evident from its synonym "*puyari*" ("*puya*" means pus, and "*ari*" means enemy). Another term *krimighna* explains the efficiency of the drug to kill worms and infections, yet another synonym "*picumarda*" is related to its potency to cure skin disease. In the Caraka Samhita, one can find the application of Neem in various clinical conditions related to skin diseases, diabetes, etc. (Trikamji 1992).

It is interesting to note that certain activities of neem mentioned by Caraka are not incorporated by other experts, e.g., anti-inflammatory effects of neem have been given more importance by Susruta while Caraka gave more importance to skin-related diseases and metabolic disorders. Susruta recommends the use of neem in the treatment of *nadivrana* (Ulcerative vascular diseases) and he advocates neem as the best drug for *danta-dhavana* (tooth brush) Susruta (Sharma 2004).

1.2 Vernacular Names of Neem

Assamese: Mahanim
Bengali: Nim, Nimgach
Gujrati: Limba, Limbado, Limado, Kohumba
Hindi: Neem, nimb, ninb, balnimb, nimgachh
Kannada: Bevu, bevina, mara, kahi, bevu, kaypebivu, olle-bevu
Malayalam: Veppu, aryaveppu, aria-bepou, ariya-veppa, aruveppu, arytikta, kaippanveppu, nimbam, pisumarddam, rajaveppu, veppa, veppuu
Marathi: Nimbay, balantanimba, kadukhajur, limb, limba, limbachajhada, nib
Oriya: Nimba
Punjabi: Nimba, Bakan, Nim
Persian: Azad-darakhte-hind, azaddarachte-hind, neeb
Tamil: Vembu, veppam, veppamaram, veppu, acutakimaram, akaluti, akappalamakkiyacatti, akuluti, ammapattini, ammapattiri, aracankanni, aricu, aristakam, aritam, arkkapatavam, arukkapatavam, arulaci, arulundi, kaduppagai, arulupati, aruluruti, arunati, aruttakam, aruttam, atipam, cakarakam, cakatam, cakatamaram, cakatamuli, cankumaru, cankumarutam, cankumarutamaram, carutopattiri, carvacatakam,

catapalacitti, catapalacittimaram, cavamuli, cenkumaru, cippuratimuli, cirilipannan, cirinapannam, cirinapanni, cirinapannimaram, cirinapattiram, cirnaparam, civam, civamatukam, civamatukamaram, iravippiriyam, kacappi, kacappu, kacappuppacitam, kacappuppacitamaram, kacappuvaruti, kacappuvarutimaram, kaitariyam, katippakai, kecamutti, kinci, kincika, kinji, kiruminacamaram, kosaram, kotakapaciyam, kotakapaciyamaram, kotaravali, kotaravalimaram, malakai, malakam, malugam, maturakkacappi, mutikam, nalatampumaram, nattuvempu, nim, nimpakam, nimpamaram, nimpataru, ninpam, niriyacam, niriyasam, nitarpam, niyamanam, niyamanamaram, niyaratam, niyatam, pacumantam, paripattiram, pariyam, parvatam, peranimpam, perunimpam, picacappiriyam, picaram, picavappiriyam, picumantam, picumattam, pirapattiram, pisidam, piyacukam, puyari, puyarikam, sengumaru, tittai, tuttai, ukkirakantam, ukkirakanti, ukkragandam, varuttam, vempumaram, venipam, vepa, veppan, vicimikini, vicumantam, vicumikini, vicumini, viruttamaram, visapatcani

Telugu: Vepa, nimbamu, vemu, yapa, taruka, vepa-chettu, yeppa

Tibetan: Ni mbaI, ni-mba

Urdu: Neem, burg neem, burge neem, gul neem, maghztukhm neem, maghz tukhm-e-neem, neem ke khusk pattay, neem ki namontian, poast darakht neem, poast darakht nim, poast neem, roghan neem, roghan nim (INMEDPLAN 2018).

1.3 Local Health Practices Related to Neem in Different Parts of India

In case of jaundice, leaf extract has been taken for jaundice by Yeravas (a tribe in Malemahadeshwara betta region of South Karnataka and Tamil Nadu)—one teacup early morning on an empty stomach for three days. Similarly, in case of rheumatism, seed oil used to massage rheumatic joints by Jenukurubas (a tribe in Malemahadeshwara betta region of South Karnataka and Tamil Nadu) (Rajendra et al. 2007).

According to the tribal people of Rajasthan, the mature fruits and young leaves are eaten widely to purify the blood. The Meena tribals residing near Sariska Tiger Reserve (Alwar) chew the fresh leaves against snake-bite and bleeding from nose (INMEDPLAN 2018).

Leaf-paste is applied to the wound of snake-bite and scorpion-sting. The seed oil has found a reputable place in birth control. Females apply a few drops of seed oil in to vagina before sexual contact. A poultice of leaves of neem, *Capparis sepiaria* (Jal) and *Tamarindus indica* (Imli) is applied on eyes by Bhils community to cure night-blindness (INMEDPLAN 2018). The seeds in Barmer and leaves in Banswara are eaten to cure abdominal pain. The bark-powder is given orally by the tribals of Udaipur region in Rajasthan to cure cough and cold, and with that of *Moringa oleifera* bark against dog-bite. Neem leaves are heated and tied over limbs for easy expulsion of guinea-worm. Paste of the flowers is ingested to cure malarial fever by the tribals of Dungarpur district (Singh and Pandey 1998). Neem leaves used as poultice for boils; fruits are preventive against, and cure for chickenpox (Panigrahi and Murti 1989). A paste made out of neem leaves and juice are useful in treating fever, wounds, and boils (CCRA&S 1990).

Tribal group Lodhas consume about 10 mL of root bark decoction against fever. They apply seed oil in the treatment of skin diseases like scabies and apply the oil on the head to promote hair growth. Santals prescribe dry flower as anthelmintic. They use leaf decoction for washing septic wounds. Oraons give stem bark extract about 20 mL to women after menstruation period as a contraceptive. Mundas use leaf extract 15 mL as febrifuge and as blood purifier repellent (Pal and Jain 1998).

The tribals of the Himalayan region use Neem as a valuable medicine as an antiseptic, used in the treatment of chickenpox. Small twigs are used as toothbrushes and as a prophylactic for mouth and teeth complaints. Extract from the leaves is useful for sores, eczema, and skin diseases. Boiled and smashed leaves serve as an excellent antiseptic. Decoction of leaves is used for purifying the blood by drinking it. This is used as a febrifuge (Biswas and Chopra 1982).

In Nepal region, tribal people of the district use the oil as hair oil and in skin diseases (Singh and Jain 2003).

In Ethno-veterinary medicine, Neem is used extensively, fumes of burned leaves are used in poultry and cattle shed as mosquito repellant (Pal and Jain 1998).

1.4 Sanskrit Synonyms of Neem

The polynomial nomenclature system used in Ayurveda has documented around 30 names over a period of time in Sanskrit denoting its morphological characters, cultural and medical uses.

1. Agnidharmava—That which reduces fire, pitta (Pra, 7/8—RJN)
2. Arista—Eradicate diseases and death (Gudu, 93—BPN, 37—MPN, 7/8—RJN)
3. Aristaphala—Fruit that which eradicate diseases (Pra, 7/8—RJN)
4. Cchardana—Induces vomiting (Pra, 7/8—RJN)
5. Devadatta—Given by the gods (Abha, 37—MPN, 7/8—RJN)
6. Hinguniryasa—Exudes gum like asafetida from the tree bark (Gudu, 93—BPN, Pra, 7/8—RJN)
7. Jyetstamalakah—That which eradicate diseases (Pra, 7/8—RJN)
8. Kakaphalah—Ripened fruits are eaten by crows (Pra, 7/8—RJN)
9. Kireta—Leaves are Adorned or used as a thorn for worshiping idols (Pra, 7/8—RJN)
10. Kusthaha—That which cure skin diseases (Abha, 37—MPN)
11. Neta—Leader, king of herbals (Gudu, 28—DHN, Abha, 37—MPN, Pra, 7/8—RJN)
12. Nimba—That which brings health and wellness (Gudu, 28—DHN, Abha, 37—MPN, Pra, 7/8—RJN)
13. Niyamana—That which pacifies diseases (Gudu, 28 DHN, Abha, 37—MPN)
14. Paribhadra—That which brings blessings and safety for all (Gudu, 93—BPN, Abha, 37—MPN, Gudu, 28—DHN, Pra, 7/8—RJN)
15. Pavanestah—That which purifies vata/brings pure air and surroundings (Pra, 7/8—RJN)
16. Picumanda—That which destroys skin diseases (Gudu, 93—BPN, Gudu, 28—DHN, Abha, 37—MPN)
17. Picumandah—That which destroys skin diseases (Pra, 7/8—RJN)
18. Picumarda—That which destroys skin diseases (Gudu, 93—BPN)
19. Pitasarakah—Exudes yellow exudates (Pra, 7/8—RJN)
20. Prabhadra—Most safe (Gudu, 28—DHN, Pra, 7/8—RJN)
21. Prabhadraka—Highly safe for all (Abha, 37—MPN)
22. Ravisannibha—Blesses all like sun rays (Abha, 37—MPN)
23. Sarvatobhadra—Safe for all (Gudu, 28—DHN, Pra, 7/8—RJN, Abha, 37—MPN)
24. Sita—Having cold potency (Pra, 7/8—RJN)
25. Sutikta—Having a high bitter taste (Abha, 37—MPN)
26. Sutiktakah—Highly bitter (Gudu, 28—DHN)
27. Tiktakah—Bitter in taste (Gudu, 93—BPN)
28. Varatikta—Best bitter medicine (Pra, 7/8—RJN)
29. Visirnaparna—That which sheds leaves/ (Pra, 7/8—RJN)

[*RJN*—Raja nighantu, *BPN*—Bhavaprakasa nighantu, *DHN*—Dhanvantari nighantu, *MPN*—Madanapala nighantu, *Pra*—Prakirna varga, *Gudu*—Guducyadi varga, *Abha*—Abhayadi varga].

1.5 Categorization in Ayurveda

Charaka Samhita: *Kandughna* (group of herbs that are useful in skin disorders) *Tiktaskandha* (group of bitter-tasting herbs)

Sushruta Samhita:

Aragwadhadi (group of fruit starting with of Cassia fistula)—This group is anti-poisonous, antipyretic, antiemetic, antiseptic, anti dermatotic
Guduchyadi (group starting with *Tinospora cordifolia*)—This group is antipyretic, appetizer, refrigerant (Sastry 2005)
Lakshadi (group of laksha)—This group is astringent, bitter, anthelminthic, antiseptic, anti dermatitis.

1.6 Major Pharmacological Properties of Neem According to Ayurveda

Taste (rasa)—Bitter (tikta) and astringent (kashaya)
Biophysical properties (guna)—Light to digest (Laghu), Dry (rooksha)
Potency (veerya)—Cold in nature (sheeta)
Post digestive effect (vipaka)—Undergoes taste conversion into pungency after digestion (katu vipaka)
Part used: Root bark, stem bark, gum, flower, leaves, seeds, seed oil (Sharma 2007)
Uses and qualities

निम्बः शीतोलघुः ग्राहीकटुतिक्तोऽग्निवातकृत्।
अहृद्यश्रमतृट्कासज्वरअरुचिकृमिप्रणुत्॥
व्रणपित्तकफछर्दिकुष्ठहृल्लासमेहनुत्। (भावप्रकाश 15th CE)

Sheeta—Imparts cooling qualities to the body
Laghu—Undergoes digestion and absorption pretty easily and quickly
Grahi—Helps in absorbing moisture from the intestine. Dries up and cleans up the moisture in wounds and ulcers
Katu—Pungent in taste; It also undergoes pungent taste conversion after digestion
Tikta—Bitter in taste
Agnikrut—Improves digestion process
Vatakrut—Causes vitiation of Vata dosha
Ahrudya—Not cordial for the heart
Shrama hara—Relieves tiredness
Trut hara—Relieves excessive thirst
Kasa hara—Helps to relieve cough
Jwara hara—Useful in fever
Aruchi hara—Helps to relieve anorexia
Krumi hara—Relieves worms; nimba is useful in intestinal worms
Vrana hara—Helps to cleanse and heal wound quickly
Pitta Kaphahara—Balances Pitta and Kapha
Chardi hrillasa hara—Helps to relieve nausea and vomiting
Kushtahara—Useful in various skin diseases
Mehanut—Useful in diabetes and disorders related to urinary tract.

1.6.1 Neem Leaf Benefits

निम्बपत्रंस्मृतंनेत्र्यंकृमिपित्तविषप्रणुत्।
वातलंकटुपाकंचसर्वअरोचककुष्ठनुत्॥ (भावप्रकाश)

Netrya—Beneficial to the eyes; helps to relieve infection
Kruminut—Helps relieve worms and microbes
Pittanut—Balances Pitta
Vishanut—Natural detoxifier
Vatala—Causes vitiation of vata
Katupaka—Pungent taste after digestion
Arochaka—Relieves anorexia
Kushtanut—Relieves skin diseases.

1.6.2 Properties of Neem Fruit

निम्बफलंरसेतिक्तंपाकेतुकटुभेदनम्।
स्निग्धंलघुउष्णंकुष्ठघ्नंगुल्मार्शकृमिमेहनुत्॥ (भावप्रकाश)

Bhedana—Helps to pass bowels easily
Snigdha—Unctuous, oily
Laghu—Light to digest
Ushna—Hot in potency
Gulmanut—Relieves bloating
Arshanut—Relieves piles (hemorrhoids)
Kriminut—Relieves worms and infection
Mehanut—Helps in diabetes
(Misra 2002).

1.6.3 Properties of Different Parts Used

Nimba pravala (new leaf) properties:

Grahi—Helps in absorbing moisture from the intestine. Dries up and cleans up the moisture in wounds and ulcers
Vata vardhaka—Causes vitiation of Vata dosha
Kapha Hara—Alleviates vitiation of Kapha dosha
Kusthahara—Useful in various skin diseases
Raktapitta hara—Relieves spontaneous hemorrhage from the mouth or nose
Krimihara—Relieves worms and infection
Netra roga hara—Helps to relieve eye disorders

1.6.4 Nimba Puspa (Flower) Properties

Pitta samaka—Pacifies vitiation of pitta dosha
Vata vardhaka—Causes vitiation of Vata dosha
Katu paka—Undergoes pungent taste conversion after digestion
Arucihara—Relieves anorexia
Krimi hara—Relieves worms and infection
Visa hara—Natural detoxifier

Nimba majja (pulp) benefits:

Krimi hara—Relieves worms and infection
Kustha hara—Useful in various skin diseases.

1.6.5 Nimba Pakva Phala (Ripe Fruit) Properties

Madhura, Tikta rasa—Sweet and bitter in taste
Snigdha—Unctuous, Guru—Heavy, Picchila—Slimy
Kapha Hara—Alleviates vitiation of Kapha dosha
Raktapitta hara—Relieves spontaneous hemorrhage from the mouth or nose
Netra roga hara—Helps to relieve eye disorders
Ksata hara—Helps in wound healing
Ksaya hara—Helps in nourishment
(Sharma 1975).

निम्बस्तिक्तरसः शीतो लघुः श्लेष्मास्रपित्तनुत् ।
कुष्ठकण्डूव्रणान् हन्ति लेपहारादिशीलितः ॥

अपक्वं पाययेच्छोफं व्रणं पक्वं विशोधयेत् ।

नात्युष्णं निम्बजं तैलं कृमिकुष्ठकफापहम् ॥
वातरक्तप्रशमनं मदालक्ष्मीज्वरापहम् । (Dhn)

Nimba is tikta in rasa (bitter in taste), laghu in guna (light in property), sita in veerya (cold in potency), alleviates slesma (vitiation of kapha dosha), asra (vitiated blood), pitta (vitiation of pitta), cures kusta (skin disorders), kandu (itching), vrana (wounds), is effective as lepa (paste). Apakva nimba alleviates sopha (unripe nimba alleviates oedema). Pakva nimba is vrana sodhaka (ripe nimba helps in wound cleansing). Nimba taila (oil of nimba) is not atyushna (neither hot nor cold in potency), pacifies vatarakta (gout) and cures krmi (worm infestation), kustha (skin disorders), kapha (vitiation of kapha dosha), mada (intoxication), jvara (fever) and alakshmi (inauspiciousness) (Kamat and Kamat 2002).

According to Raja nighantu,

"प्रभद्रकः प्रभवति शीततिक्तकः
कफव्रणकृमिवमिशोफशान्तये ।
वलासभिद् बहुविषपित्तदोषजिद्
विशेषतो हृदयविदाहशान्तिकृत् ॥
निम्बतैलं तु नात्युष्णं कृमिकुष्ठकफापहम् । " (रा.नि.)

Neem is cold in potency, it has bitter taste and has great potential in curing diseases of kapha, ulcers (vrna), worms (krimi), vomiting (vami), and swelling (sopha). It also reduces the ill effects of poisonous bites and vitiation of pitta. It has a special potency to treat cardiac ailments due to inflammation in general. Its seed oil is not too hot and highly beneficial in worms and kapha predominant skin disorders (Tripathi 1982).

Flowers: Nimba Vrkasaya Puspa (flowers of nimba) is Pittaghna (alleviates vitiation of pitta) specifically. It is Tikta in Rasa (bitter in taste), Krmighna (cures worm infestation), and Kapha hara (alleviates vitiation of kapha dosha) (Fig. 1.1).

Tender twigs: Nimbasya Suksma Sakha (tender twigs) cures Kasa (cough), Svasa (dyspnoea), Arsa (piles), Gulma (abdominal tumors), Krmi (worm infestation), Meha (diabetes).

Fruits: Its Ama Phala (ripe fruits) is Laghu (light in property), Snigdha (unctuous), Bhedana (piercing or cutting in nature), Usna (hot) in Potency, cures Meha (diabetes), Kustha (skin disorders). It is Tikta in Rasa (bitter in taste), Katu in Paka (pungent in taste post digestion), cures Gulma (abdominal distention/tumors), Arsa (piles), Krmi (worm infestation) and Meha (diabetes) (Fig. 1.2).

Seed marrow: Nimba Bijasya Majja (seed marrow) is Kusthaghna (cures skin disorders), Krmi nasini (cures worm infestation) (Fig. 1.3).

Oil: Nimba Taila (oil of nimba seeds) is Kusthaghna (cures skin disorders), Tikta in Rasa (bitter in taste), Krmi hara para (best to cure

"निम्बवृक्षस्य पुष्पाणि पित्तघ्नानि विशेषतः ।
तिक्तानि च कृमिघ्नानि तथा कफहराणि च ॥
निम्बस्य सूक्ष्मशाखा तु कासश्वासार्शःगुल्महा ।
कृमिमेहहरा प्रोक्ता फलं चामं लघु स्मृतम् ॥
स्निग्धं च भेदकं चोष्णं मेहकुष्ठविनाशकम् ।
आमं फलं रसे तिक्तं पाके तु कटुकं मतम् ॥
स्निग्धं लघूष्णं कुष्ठघ्नं गुल्मार्शःकृमिमेहनुत् ।
निम्बबीजस्य मज्जा तु कुष्ठघ्नी कृमिनाशिनी ॥
निम्बतैलन्तु कुष्ठघ्नं तिक्तं कृमिहरं परम् ।
निम्बवृक्षस्य पञ्चाङ्गं रक्तदोषहरं मतम् ॥
पित्तं कण्डूं व्रणं दाहं कुष्ठं चैव विनाशयेत् । (शा. नि.)

Fig. 1.1 *Nimba Vrkasaya Puspa*/Flowering branches of Neem

Fig. 1.2 Fruits of Neem

Fig. 1.3 Nimba Bijasya Majja

worm infestation). Nimba Vrksasya Panchanga (all parts of niba) is Rakta dosha hara (alleviates vitiation of blood). It cures Pitta (vitiation of pitta), Kandu (itching), Vrana (wounds), Daha (burning sensation) and Kustha (skin disorders) (Shaaligramavaishya 1997).

निम्बवृक्षो लघुः शीतस्तिक्तो ग्राही कटुः स्मृतः ।
अग्निमान्द्यकरश्चैव व्रणशोधनकारकः ॥
शोफपाककरो बाले हितो रुद्यो मतो बुधैः ।
कृमिवान्तिव्रणकफशोफपित्तविषापहः ॥
वातं कुष्ठं च हृद्दाहं श्रमं कासं ज्वरं तृषां ।
अरुचिं रक्तदोषं च मेहं चैव विनाशयेत् ॥
कोमलः पल्लवश्चास्य ग्राहको वातकारकः ।
रक्तपित्तं नेत्ररोगं कुष्ठं चैव विनाशयेत् ॥
जीर्णपर्णं विशेषेण व्रणनाशकरं मतम् ।" (नि.र.)

Nimba Vrksa (tree) is Laghu (light in property), Sita in virya (cold in potency), Tikta, Katu, in rasa (bitter and pungent in taste), Grahi (water absorbent), Agnimandyakara (decreases digestion), Vranasodhanakara (wound healing). Cures Krmi (worm infestation), Vanti (vomiting), Vrana (wounds), Kapha (alleviates vitiation of kapha dosha), Sopha (oedema), Pitta (alleviates vitiation of pitta dosha), Visa (poisoning), Vata, Kustha (skin disorders), Hrddaha (burning sensation in the chest), Srama (exhaustion), Kasa (cough), Jvara (fever), Trsa (morbid thirst), Aruci (tastelessness), Rakta dosha (alleviates vitiation of blood), Meha (diabetes).

Tender leaves: Komala Pallava (tender leaves) is Grahi (water absorbent), Vata kara (causes increase in vata dosha), cures Raktapitta (hemorrhagic disorders), Netra roga (eye disorders), Kustha (skin disorders).

Matured leaves: Jirna Pallava (matured leaves) are Vrana nasa kara (heals wounds) specifically (Nighantu ratnakara).

"निम्बस्तिक्तः कटुः पाके लघुः शीतोऽग्निवातकृत् ।
ग्राह्यह्रद्यो जयेत् पित्तकफमेहज्वरकृमीन् ॥
कुष्ठकासारुचिश्वासहृल्लासश्वयथुव्रणान् ।
ग्राहि प्रवाळं निम्बस्य रक्तपित्तकफकृमीन् ॥
कुष्ठघ्नं वातजननं नेत्ररोगान् विनाशयेत् ।
तद्वत् पत्राणि निम्बस्य व्रणघ्नानि विशेषतः ॥
शलाका निम्बपत्रस्य कासश्वासविनाशिनी ।
कृमिघ्ना तु वरिष्ठा स्यात् कुष्ठज्वरविनाशिनी ॥
चक्षुष्यं निम्बपुष्पं च कृमिपित्तविषप्रणुत् ।
वातळं कटुपाकं स्यात् सर्वारोचकनाशनम् ॥
फलं तिक्तं रसे पाके कटुकं भेदनं लघु ।
अरूक्षमुष्णं कुष्ठघ्नं गुल्मार्शःकृमिमेहनुत् ॥
निम्बस्य पक्वं मधुरं सतिक्तं
स्निग्धं फलं शोणितपित्तरोगे ।
कफे प्रशस्तं नयनामयघ्नं
क्षतक्षयघ्नं गुरु पिच्छिलं च ॥
निम्बबीजस्य मज्जा च कृमिकुष्ठविशोधनः ।
नात्युष्णं निम्बजं तिक्तं कृमिकुष्ठकफप्रणुत् ॥
अभ्यङ्गान्नावनात् क्षीरभोजिनः पलितापहम् ।" (कै.नि.)

Nimba is tikta in rasa (bitter in taste), laghu in guna (light in property), sita in veerya (cold in potency), katu in paka (pungent post digestion), increases agni and vata (increases digestion and causes vitiation of vata dosha), is grahya (water absorbant), hrdya (cordial to the heart), cures pitta kapha (vitiation of pitta and kapha dosha), meha (diabetes), jvara (fever), krmi (worm infestation), kusta (skin disorders), kasa (cough), aruci (tastelessness), svasa (dyspnoea), hrllasa (nausea), svayathu (oedema) and vrana (wounds).

Nimba pravala (tender leaves) is grahi (water absorbant), increases vata (causes vitiation of vata dosha), cures rakta pitta (hemorrhagic disorders), kapha (vitiation of kapha dosha), krmi (worm infestation), kusta (skin disorders) and netra roga (eye disorders). Its patra (leaves) alleviates vrana (wounds).

Nimba puspa (flowers of nimba) is katu in paka (pungest post digestion), caksushya (beneficial to the eyes), increases vata (causes vitiation of vata dosha), alleviates kapha, pitta (vitiation of kapha and pitta dosha), visa (poisoning), and sarva arocaka (all types of taste disorders).

Phala (fruit) is tikta in rasa (bittet in taste), usna in virya (hot in potency), laghu in guna (light in property), katu in paka (pungent post digestion), bhedana (piercing in nature), cures kustha (skin disorders), gulma (abdominal tumors), arsa (piles), krmi (worm infestation), and meha (diabetes).

Pakva phala (ripe fruit) is madhura, tikta in rasa (sweet and bitter in taste), snigdha (unctuous), guru (heavy), picchila in guna (slimy in quality), cures sonita pitta roga (hemorrhagic disorders), kapha (vitiation of kapha dosha), nayana Amaya (eye disorders), ksata (wounds), and kshaya (emaciation).

Nimba bijasya majja is krmi kustha visodhana (Seed marrow cures worm infestation and skin disorders) (Sharma 1975).

1.7 Major Therapeutic Uses

Urticaria: Neem leaves and Amalaki dried fruit are mixed with ghee and given regularly (Sharma 2007).
Diabetes: Decoction of nimba is a specific remedy (Sharma 2004).
Fever: The leaves, root, fruit, and bark of neem is mixed with ghee and fumigation is given (Conrick 2001).
Skin diseases: Decoction of nimba and patola is efficacious in skin diseases taken in a dose of 40–60 mL (Trikamji 1992).

Intake of nimba and amalaka for a month overcomes all types of skin diseases (Sharma 2007).

Local application of the juice of nimba destroys skin diseases such as eczema, ringworms etc. (Srikanthamurthy 2006).

Wounds: Decoction of nimba can be used to clean wounds (Trikamji 1992).

Nimba leaf mixed with honey acts as a cleansing agent. Both of them added with ghee promotes healing (Sharma 2004).

Paste of nimba leaves and sesamum mixed with honey cleanses wounds while mixed with ghee acts as a healing agent (Sharma 2007).

Paste of nimba leaves on external application cleanses and heals wounds while its intake alleviates vomiting, skin disorder and worms (Srikanthamurthy 2006).

1.8 Compound Formulations

There are around 300 compound formulations in Ayurveda in which Nimba is a major ingredient (INMEDPLAN 2018).

The important compound formulations are (1) Mahathikthakam kashya and Ghrita, (2) Pancha thikthakam kashaya, (3) Nimbadi churna (4) Thikthakam Kashaya and grita (AFI 2000).

For example, a combination of Nimba (Neem), *Boerhavia diffusa* (varsabhu), *Pongamia glabra* (naktamala) and *Calotropis gigantea* (arka) is made into a paste and applied over affected areas, in case of localized swelling.

A decoction is prepared using Nimba (neem), *T. cordifolia* (guduci) mixed with honey and gingely oil used as gandusa (gargle) for diseases of mouth like mouth ulcer, inflammatory growths (Vaidya 1995).

Acknowledgements We thank Dr. P. M. Unnikrishnan and Dr. Subramanya Kumar K. for their valuable inputs and encouragements.

References

AFI (2000) The ayurvedic formulary of India. NISCAIR, CSIR, New Delhi
Biswas, Chopra (1982) Common medicinal plants of Darjeeling and the Sikkim Himalayas. Soni Reprints Agency, New Delhi, India
Conrick J (2001) Neem the ultimate herb. Pilgrims Publishing Varanasi
CCRA&S (1990) Glimpses of medico-botany of Bastar district. CCRA&S, New Delhi, India
INMEDPLAN (2018) Indian medicinal plants database. The University of Transdisciplinary Health Sciences and Technology (TDU), Bengaluru, Karnataka, India
Kamat DK, Kamat SD (2002) Dhanvantari nighantu. Chaukhamba Sanskrit Pratishthan, New Delhi
Misra B (ed) (2002) Bhavamisra. Bhavaprakasha nighantu. Krishnadas Academy, Varanasi
Nair (2015) Knowledge of pharmacogenomics in Indian traditional medicine—ayurveda. J Pharmacogenomics Pharmacoproteomics 6:150
Pal DC, Jain SK (1998) Tribal medicinen. Naya Prokash, Calcutta, India

Panigrahi G, Murti SK (1989) Flora of Bilaspur district, botanical survey of India, Calcutta, vol 1

Rajendra D, Kshirsagar D, Singh NP (2007). Ethnobotany of Mysore and Coorg, Karnataka State, Bishen Singh Mahendra Pal Singh, India

Sastry JLN (2005) Illustrated Dravyaguna Vijnana Chaukhambha Orientalia, Varanasi

Shaaligramavaishya (1997) Shaaligrama nighantu Bhushanam. Khemraj Sreekrishna Das, Mumbai

Sharma PV (ed) (2004) Susruta. Susruta Samhita. Published by Chaukhambha Visvabharati, Varanasi

Sharma PV (2007) Cakradatta, Chaukambha sanskrit series, Varanasi, India

Sharma PV, Sharma GP (eds and trans) (1975) Kaiyadeva. Kaiyadeva nighantu. Chaukhambha Orientalia, Varanasi

Singh V, Jain AP (2003) Ethnobotany and medicinal plants of India and Nepal, vols I & II. Scientific publishers (India), Jodhpur, India

Singh V, Pandey RP (1998) Ethonobotany of Rajasthan, India. Scientific Publishers (India), Jodhpur, India

Srikanthamurthy KR (ed) (2006) Sarngadhara. Sarngadhara Samhita. Chaukhambha Orientalia, Varanasi

Trikamji VY (ed) (1992) Charaksamhita of agnivesha, 4th edn. Munshiram, Manoharlal Publishers Pvt. Ltd., New Delhi

Tripathi I (ed) (1982) Raja Narahari. Raja nighantu. Krishnadas Academy, Varanasi

Vaidya HP (ed) (1995) Ashtangahridaya. Krishnadas Academy, Varanasi, Chaukhamba Press, Varanasi

Heritage of Neem–Peepal Tree Resides a Profound Scientific Facts

K. L. Ashalatha and Malali Gowda

Abstract

India's mythology and traditional medicines have great significance to Neem (*Azadirachta indica*) and Peepal (*Ficus religiosa*). Some of the historic values, rituals, and festival practices taken for these trees are mentioned here. After the meditation of great Indian sage Buddha, the Peepal tree is named as Bhodi tree. Neem–Peepal trees are widely used for treating many diseases in traditional medicines like Ayurveda, Homeopathy, Unani, and Siddha. In Ayurveda, few of the shlokas (quotes) describing the medicinal properties of Neem–Peepal trees are given in this chapter.

2.1 Introduction

The divine trees of India, Neem (*Azadirachta indica*) belongs to Meliaceae family and Peepal or Bodhi tree (*Ficus religiosa*) belongs to Moraceae family. There are several vernacular names of Neem: Nimba, Bevina mara, etc. and Peepal are Ashwatta, Arali, Bodhi, etc. These tree species are considered as keystone and holy in Indian subcontinent. These are semi-evergreen, terrestrial, and adopt to xerophyte nature. Since the ancient days to contemporary world, these trees are believed to be sacred and worshiped by various communities like Hindus, Buddhists, Jains, etc.

2.2 Ashwattakatte or Aralikatte Platform

We could encounter the Neem and Peepal trees that are grown together near Indian temples and villages. The **Ashwattakatte** or **Aralikatte** platform is often constructed using rocks around these trees (Fig. 2.1). In the early dawn of morning, devotees do pradakshina (circumambulation) and prayer for these trees (Fig. 2.2).

In southern parts of India, people perform marriage between Neem and Peepal trees which are planted and grown together. Hence they coin Neem as female and Peepal as male. The above image depicts the marriage of Neem and Peepal tree (Fig. 2.1). The white-colored cloth (Panche—in Kannada language) been tied to the left side Peepal tree and considered it as male (groom) and the Indian traditional saree is been tied to the middle Neem tree, as female (bride) (Fig. 2.1) and also we can witness the Nagas idols (Snake sculpture) are placed under these trees. Later, the traditional marriage rituals were made for these trees.

K. L. Ashalatha
Centre for Functional Genomics and Bioinformatics, The University of TransDisciplinary Health Sciences and Technology, Bengaluru, India

M. Gowda (✉)
The University of TransDisciplinary Health Sciences and Technology, Bengaluru, India
e-mail: malalig@tdu.edu.in

Fig. 2.1 Ashwatta or Aralikatte platform near India's temple. *Credit* Ashalatha K. L.

2.3 Neem is Symbolic to Mariamman Temple

Neem is associated with Indian culture. For example, Mariamman festival celebrated across the Tamil Nadu state during the Adi or April month. Mari is called as Goddess Parvathi, Durge, Kali, Shitaladevi, etc. (Fig. 2.3).

Thousands of people gather during Mariamman festival and carry neem leaves from one house to another for worshiping goddess Mari. People walk for miles, carrying water mixed with neem leaves and turmeric powder to ward-off illnesses or diseases like measles, cholera, smallpox, and chickenpox. The outbreak of these diseases is so high during summer period. There is a logical and scientific thorough process discovered by ancient people in India and converted these ideas into traditional and cultural practices or festival.

2.4 Bevu-Bella During Ugadi Festival

In Karnataka and Andhra Pradesh, people celebrate the Ugadi festival during March–April (Chaitra Masa). This is considered as the New year for local people. On Ugadi day, people consume Neem and Jaggery (Bevu-Bella) together which signifies that life is formed with happiness and sorrows or like the taste of bitter and sweet. They also add Neem leaves into the hot water and take bath on that day which is good for skin health. This fest is celebrated among the Hindu families (Fig. 2.4).

Fig. 2.2 Devotees circumambulation to the Aralikatte. *Credit* Dr. Malali Gowda

Fig. 2.3 Use of Neem leaves during Mariamman festival. *Credit* Kamala

Fig. 2.4 Neem and jaggery (Bevu-Bella) consumed on Ugadi festival. *Credit* Shreyas

Thus, this concept needs to be translated among the local communities to create an awareness of scientific importance of this festival. Neem has several properties like antimicrobial, antiviral, antifungal, and antidiabetic properties and also widely used as a pesticide in agriculture (Brahmachari et al. 2004). The different tissues of Neem like twig, leaves, roots, bark, flowers, fruits, seeds, and bark have been used for preparation of traditional medicines and bio-pesticides (Brahmachari et al. 2004). Further, we have sequenced the Neem genome to understand the metabolic pathways (Kuravadi et al. 2015).

Peepal tree is believed to harbor God Trimurtis (Lord Shiva, Lord Vishnu, and Lord Brahma) in Hinduism (mentioned in ancient traditional texts). The Buddhist believes that Buddha attained enlightenment underneath the Peepal tree. After his meditation, this tree became largely popularized as "**Bodhi tree**". The Mahabodhi Temple is located in Bodh Gaya in Bihar, India is known as Buddhist pilgrimage in the world. Bodhi tree at Bodh Gaya temple was the offshoot of the original propagule planted in 288 BCE (CABI 2018) (Fig. 2.5).

Other than religious beliefs and usage, these trees are privileged to have scientific facts. Most of the traditional medicines such as Ayurveda, Siddha, and Homeopathy describe the usage of various tissues (leaf, bark, roots, fruits, twigs, etc.) of Neem and Peepal for making medicines and formulations.

Peepal tree has various medicinal properties including antioxidants, antimicrobial, anti-acetylcholinesterase, wound healing, antidiabetic, anti-inflammatory, etc. (Gautam et al. 2014). It is used for many medicinal purposes such as diabetes, ulcers, gastrointestinal problems, neurological disorders, skin diseases, urinary disorders, respiratory problems, etc. Some of the Ayurvedic formulations, which use Peepal tissues, are like Nalpamardi thailam, Nyagrodhadi churna, Sarivadyasava, Panchavalkadi tailam. Recently, we have sequenced genome of Peepal tree to decode the biomedical properties (*unpublished* data) (Fig. 2.6).

Behind every traditional practice, there is a hidden scientific knowledge, yet to explore, observe, understand, and narrate it carefully. Concept of Aralikatte structure is an interesting biological knowledge where roots are trained towards the ground. If this structure is not constructed, then roots of Peepal tree pierce aerially and damage houses. Thus, Aralikatte structure is covered with rocky material, which guide the Peepal roots towards the underground soil. It is the best example where traditional knowledge that has scientific thought process.

Some of the useful properties of Neem and Peepal tree are been described in the Ayurveda texts. And few of those shlokas (quotes) were taken from Ayurveda listed below.

Neem Tree:

Neem Tree:

||भावप्रकाशनघिण्टु||

निम्बः शीतो लघुरग्राही कटुपाकोऽग्नवातनुत् |
अहृद्यः श्रमतङ्कासज्वरारुचकृमिपिरणुत् ||८२||
वरणपित्ततकफच्छर्दकिुष्ठहृल्लासमेहनुत् |

निम्बः शीतो - Neem tree is cold in condition
लघुरग्राही - Quick and easy to digest and absorbs the moisture from intestine. It is used treat the ulcer and heals the wounds.
कटुपाकोऽग्नवातनुत् - Having a pungent taste, bitter taste, helps in digestion, increase Vata prakriti (ayurveda classified type)
अहृद्यः - It is not good for heart
श्रमतङ्कासज्वरारुचकृमिपिरणुत् - It helps in relieving tiredness and thirst. It decreases the Pitta (ayurveda classified type) hence associated with fever and thirst. It helps in respiratory problems, relieves cough and also has antimicrobial property. Useful for relieving anorexia disorder, removes worms in intestine and heals wounds quickly. It balances the Pitta and Kapha prakriti. Helps to relieve nausea and vomiting. Useful for skin diseases, diabetes problem and urinary problems.

Source e-Nighantu (Collection of Āyurvedic Lexicons)

Peepal Tree:

Peepal tree:
|| कैयदेवनघिण्टु ||
अश्वत्थः शीतलो रूक्षः कषायो दुर्जरो गुरुः ||४३२||
वरणपित्ततकफासृग्घ्नो वरण्यो योनिविशोधनः |४३३|

अश्वत्थः शीतलो रूक्षः - Ashvattha tree has coolant property
कषायो दुर्जरो गुरुः - Having an astringent taste and difficult to digest
वरणपित्ततकफासृग्घ्नो - It is used to treat the ulcers and wounds. And also balances the pitta, kapha doshas [types of prakriti (nature) classified in ayurveda]
वरण्यो - It gives the good complexion to skin and improves the skin color.
योनिविशोधनः - It also cures the female reproductive diseases

||राजनघिण्टु||
पिप्पलः सुमधुरस्तु कषायः शीतलश्च कफपित्तवनाशी |
रक्तदाहशमनः स हि सद्यो योनिदोषहरणः किल पक्वः ||११४||
अश्वत्थवृक्षस्य फलानि पक्वान्यतीवहृद्यानि च शीतलानि |
कुर्वन्ति पित्तासृविषार्तिदाहविच्छर्दिशोषारुचिदोषनाशम् ||११५||
सुमधुरस्तु - sweet in taste
रक्तदाहशमनः - Useful for bleeding disorder
सद्यो योनिदोषहरणः - Quick relieve from vaginal and urinary tract problems
किल पक्वः - Bark is used
फलानि पक्वान्यतीवहृद्यानि - Fruits are tasty and cold
पित्तासृविषार्तिदाहविच्छर्दिशोषारुचिदोषनाशम् - Removes toxic and poisonous condition. And it relieves burning, nausea, weakness anorexia and cures skin diseases

Source e-Nighantu (Collection of Āyurvedic Lexicons)

Fig. 2.5 Buddha meditating under the Peepal tree and circumambulation by devotees. *Credit* Ashalatha K. L.

Fig. 2.6 **a** Aralikatte built around the Peepal using rock and **b** showing roots of Peepal tree. *Credit* Sachin and Rakshith

Acknowledgements I would like to thank Dr. Malali Gowda for giving an opportunity to contribute in this book and his motivation for writing this article. I would like to thank Mr. Shreyas, Ms. Kamala, Mr. Sachin, and Mr. Rakshith for contributing the photographs, which are used in this chapter.

References

Brahmachari G et al (2004) Neem—an omnipotent plant: a retrospection. ChemBioChem 5:408–421

CABI (2018) *Ficus religiosa* (sacred fig tree). Retrieved 23 July 2018. Invasive Species Compendium, CABI

e-Nighantu (Collection of Āyurvedic Lexicons), Central Council for Research in Ayurvedic Sciences (CCRAS), New Delhi. http://niimh.nic.in/ebooks/e-Nighantu/

Gautam S, Meshram A, Bhagyawant SS, Srivastava N (2014) *Ficus religiosa*—potential role in pharmaceuticals. Int J Pharm Sci Res 5(5):1616

Kuravadi NA, Yenagi V, Rangiah K, Mahesh HB, Rajamani A, Shirke MD, Russiachand H, Loganathan RM, Shankara Lingu C, Siddappa S, Ramamurthy A, Sathyanarayana BN, Gowda M (2015) Comprehensive analyses of genomes, transcriptomes and metabolites of neem tree. PeerJ 3:e1066

Method to Quantify Plant Secondary Metabolites: Quantification of Neem Metabolites from Leaf, Bark, and Seed Extracts as an Example

Kannan Rangiah and Malali Gowda

Abstract

Among the medicinal plants, neem has its own value in terms of treating many known and unknown diseases. Neem plant is known to contain several thousands of secondary metabolites, which are crucial for multifunctional properties like anti-oxidation, anti-inflammation, antimalarial, and anticarcinogenic activities. Now it is important to understand the molecular details like exact quantity of the neem metabolites in different parts of the plants. Here we have showed a UHPLC-MS/SRM method to quantify five neem metabolites (Azadirachtin, Nimbin, Salanin, Azadiradione, Epoxy/Hydroxy-azadiradione) from different parts of neem plants (leaf, bark, and seed). Among the five metabolites analyzed, E/H-Azadi is present in very high concentration in neem plant (leaf: 124,239 pg/μg, bark: 906.97 pg/μg, seed: 7309.48 pg/μg) as compared to other metabolites. Interestingly, E/H-Azadi seems to be the most abundant metabolite in the neem leaf and bark extracts and azadi is the highest in the seed extract. In the leaf extract, E/H-Azadi is ~136 fold higher compared to bark and ~17 fold higher compared to seed extract. The amount of E/H-Azadi in leaf is 124,239 pg/μg of leaf extract, which constitutes ~10% in the leaf extract. The excess amount of E/H-Azadi in the neem leaf might be one of the reasons for its multifunctional properties in nature.

Abbreviations

Aza	Azadirachitin
Azadi	Azadiradione
CID	Collision-induced dissociation
E/H-Azadi	Epoxy or hydroxyazadiradione
SRM	Selected reaction monitoring
UHPLC-MS/SRM	Ultrahigh performance liquid chromatography/mass spectrometry/selected reaction monitoring

---

K. Rangiah (✉)
Food Safety & Analytical Quality Control Laboratory, CSIR-Central Food Technological Research Institute (CFTRI), Mysore, Karnataka 570020, India
e-mail: kannan.rf0829@cftri.res.in

M. Gowda
Centre for Functional Genomics and Bioinformatics, The University of TransDisciplinary Health Sciences and Technology, Foundation for Revitalization of Local Health Traditions, Bengaluru, India

© Springer Nature Switzerland AG 2019
M. Gowda et al. (eds.), *The Neem Genome*, Compendium of Plant Genomes, https://doi.org/10.1007/978-3-030-16122-4_3

3.1 Introduction

One of the nature's gifts is "medicinal plants", which we have been using for thousands of years in daily life to treat numerous human diseases all over the world. In rural areas of the developing countries, plant-derived extracts are used as primary source of medicine. About 80% of the people in developing countries use traditional medicines for their healthcare (FAO 1997). The side effects are the major issues in the existing drugs in the market. Due to this, there is also a shift towards traditional medicine like Ayurveda and increased usage of herbal products in the recent past (Debas et al. 2006). Plants are the only major source of drugs for the majority of the world population. There are approximately 500,000 plant species occurring worldwide, of which only 1% has been phytochemically investigated, there is great potential for discovering novel bioactive compounds (Gordon and David 2013). Plants are known to contain many kinds of nutrients, including macronutrients (carbohydrates, proteins, and lipids) and micronutrients (minerals and vitamins), that shows positive effects on human health. There are non-nutritive substances have also been found in plants, such as phenols, flavonoids, alkaloids, tannins, and terpenoids, collectively known as phytochemicals which helps plants to survive in its environment (Mamta et al. 2013). The phytochemicals are known to contain multifunctional properties like anti-oxidation, anti-inflammation, cardio-protection, antimalarial, and anticarcinogenic activities (Mamta et al. 2013). One of the best examples for antioxidant properties is the extract of tea leaves (*Camellia sinensis*) due to the presence of polyphenols (like epigallocatechin gallate and epigallocatechin). It is the most widely consumed beverage in the world, next to water. Secondary metabolites from tea are also known to prevent chronic diseases, including cardiovascular diseases, cancer, and obesity due to their anti-inflammatory and anti-allergic effects (Alexandr et al. 2015). For antimalarial properties, the best-known plant extracts are from Cinchona tree (quinine) and *Artemisia annua* (artemisinin) used from many years to treat malaria (Vincent et al. 2008). More recent anticancer agent paclitaxel (more commonly known as taxol), was isolated from the bark of the Pacific yew tree (Priyadarshini and Keerthi 2012). Several other plant-derived compounds are currently in preclinical and clinical trials. There are other important plants like Amla, Ashok, Aswagandha, Brahmi, Long pepper, Sandalwood, Tulsi, Gritkumari, and Neem, where the secondary metabolites are widely used for health-related problems (Wungsem et al. 2013). Among these, neem has gained the distinction of being the most researched tree in the world; still nothing about neem is yet definite. The neem genome has been published (Nagesh et al. 2015); the metabolome of neem will help us in understanding more details of neem biology.

The neem tree was first discovered in India about 4500 years ago. The neem research was started in 1920s by Indian scientists; neem's ability to repel insects was reported in the scientific literature in 1926–1929 (National Research Council 1992). Neem is native to India and Burma, and also found in Bangladesh, Pakistan, and Nepal. It is a member of the mahogany family, Meliaceae, Genus—Azadirachta and Species—indica. It is today known by the botanic name *Azadirachta indica* A. Juss. Neem ingredients are applied in Ayurveda, Unani, Homeopathy, and modern medicine for the treatment of many infectious, metabolic, or cancer diseases (Mohammad 2016). The botanicals (plant extracts) of neem (*A. indica*) have enormous potential to influence modern agrochemical research. It has been called as "village pharmacy" and "A tree for solving global problems", due to its unique multifunctional anti-inflammatory, antipyretic, antihistamine, antifungal, antibacterial, anti-ulcer, analgesic, anti-arrhythmic, antitubercular, antimalarial, diuretic, spermicide, anti-arthritic, antiprotozoal, insect repellant, antifeedant, and antihormonal properties (Mohammad 2016). It has one of the richest sources of secondary metabolites in nature. Over 300 compounds have been isolated from different parts of the neem plant. One-third of the compounds belong to the limonoid group of natural products, which are the major cause for these widespread activities. These compounds have a low toxicity against non-target and beneficial organisms and cause less disruption to ecosystems than conventional insecticides (Ruchi et al. 2013). The information about 250 neem secondary metabolites was documented in the NeeMDB (Kaushik et al. 2014).

Limonoids, are the most important and well-studied class of triterpenoids and about one-third of phytochemicals isolated from neem belong to this class (Champagne et al. 1992; Tan and Luo 2011). Azadirone, azadiradione (Azadi), isonimolide, azadirachtin (Aza), salanin, nimbolide, gedunin and 7-deacetyl-7-benzoyle poxyazadiradione are few representative limonoids (Avinash et al. 2015). The most important active constituent is Aza and the others are nimbolinin, nimbin, nimbidin, nimbidol, sodium nimbinate, gedunin, salanin, and quercetin (Tan and Luo 2011; Avinash et al. 2015). Leaves contain ingredients such as nimbin, nimbanene, 6-desacetylnimbinene, nimbandiol, nimbolide, ascorbic acid, n-hexacosanol, and amino acid, 7-desacetyl-7-benzoylazadiradione, 7-desacetyl-7-benzoylgedunin, 17-hydroxyazadiradione (17-H-Azadi), and nimbiol (Ali 1993; Hossain et al. 2011; Kokate et al. 2010). So far, at least nine neem limonoids have demonstrated ability to block insect growth, affecting a range of species that includes some of the most deadly pests of agriculture and human health. New limonoids are still being discovered in neem, but Aza, nimbin, salanin, Azadi, and epoxyazadiradione (E-Azadi) are the best known and, for now at least, seem to be the most significant pesticidal and/or medicinal principle (Athar et al. 2012). The chemical structures of these compounds are shown in Fig. 3.1. To understand the biosynthetic pathways involved in the synthesis of Aza A in neem is still a challenging problem.

Neem based insecticides constituted about one-third of the botanicals used in agriculture as per the 2003 survey (Isman 2006). Crude extracts of neem are found to be more potent than pure Aza suggesting that there are many more compounds in the neem extract, which even at low concentration have potentiating abilities. Aqueous, methanolic, and ethanolic extracts of neem seeds show biological activity in the laboratory and in the field, although at a varying extent to different target organisms (Selma et al. 2013; Fauziah et al. 2012). Many studies are, however, limited by the extracts, which are not chemically defined. Moreover, the concentration of metabolites in neem varies with factors such as tree age, different parts, and geographical location. In most of the studies, the crude extracts of neem explants were used for checking the activity (Selma et al. 2013; Fauziah et al. 2012). There are some neem metabolites, which are commercially available; very few labs are really purifying the neem metabolites from the crude extracts of leaf, bark, and seed (Saikat et al. 2014). With the available neem metabolites, now there is chance to check the absolute quantification of these metabolites in the crude extracts with the help of advance mass spectrometry based quantitative metabolomics approaches (Kannan et al. 2016). These approaches might help us in finding out the chemical entity, which are responsible for the biological action in the neem plant extracts. In this chapter, we focus mainly on the application of metabolomics approach (targeted) to understand the level of five neem metabolites in the different parts of the tree (leaf, bark, and seed) by using advanced Ultra-high Performance Liquid Chromatography-Mass Spectrometry/Selected Reaction Monitoring (UHPLC-MS/SRM) method. This kind of method can also be applied for the quantification of secondary metabolites from other plants (based on the availability of the metabolite standards) using advanced mass spectrometry techniques.

3.2 Neem Metabolite Standard Preparation and Analysis by UHPLC-MS/SRM Method

In the recent years, Liquid Chromatography and Mass Spectrometry (LC-MS) based quantifications has considered as gold standard due to its specificity, sensitivity, and selectivity (Simon

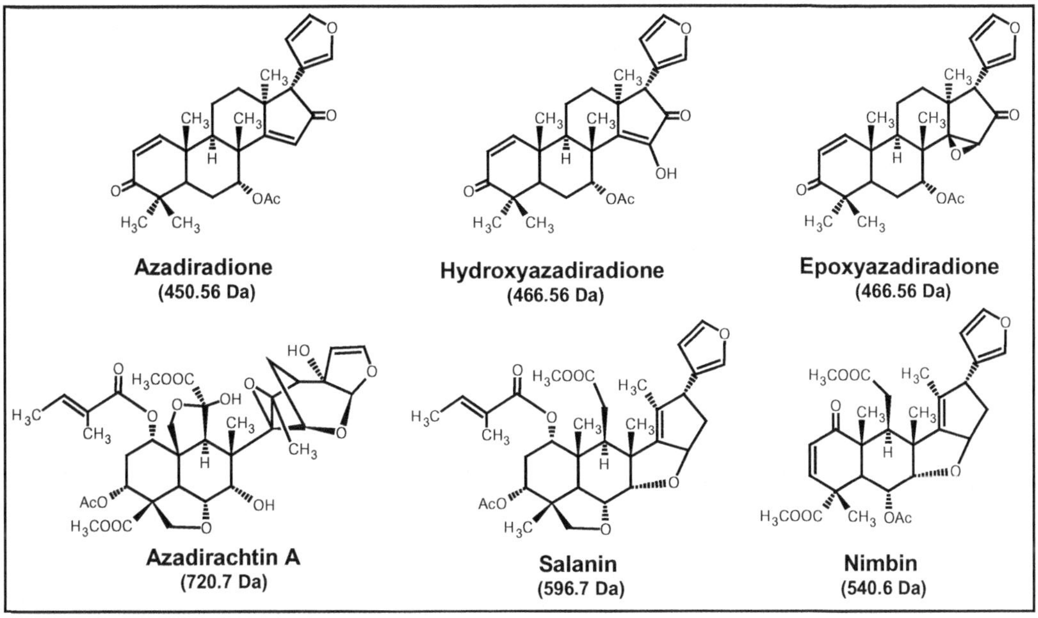

Fig. 3.1 Chemical structure of neem metabolites analyzed in this study

et al. 2014). By using UHPLC in the front end of MS showed enormous potential for separating compounds using minor particle size columns and there is also reduction in the overall running time compared to traditional LC system. In the MS system, the methods like selected reaction monitoring (SRM) or multiple reactions monitoring (MRM) showed thousand-fold higher sensitivity compared to normal full scan methods. To establish the sensitive method for the quantification of neem metabolites, we have obtained Azadirachtin (Sigma-Aldrich, Bangalore, India), Nimbin (TRC, Canada), Salanin (Chromodex, USA) and Azadiradione (Sami Labs, Bangalore, India). Epoxy/Hydroxyazadiradione was purified from neem leaf extract as per the published procedure (Kannan et al. 2016). The Estrone-d4 used as an internal standard for the assay was purchased from Steraloids Inc. (Newport, RI, USA).

The main stock (1 mg/mL) and 100-fold diluted stock (10 μg/mL) was prepared in methanol and stored at −80 °C. Further, the working solutions were prepared by mixing the required amounts of corresponding stock solutions and performing serial dilutions with methanol to the desired concentration. The highest concentrations on column were as, Aza A (5 ng), nimbin (0.4 ng), salanin (2 ng), azadi (2 ng), E/H-Azadi (2 ng). The lowest concentrations on column were as, Aza A (78 pg), nimbin (6.2 pg), salanin, azadi, E/H-Azadi were (15.6 pg). The stock of 100 μg/mL of estrone-d4 was prepared in methanol from the main stock (1 mg/mL) and stored at 4 °C. The UHPLC-MS conditions were quite similar to those used in our previous studies (Kannan et al. 2016; Kannan 2014; Nivedita et al. 2015). A TSQ Vantage triple quad instrument (Thermo Fisher Scientific, San Jose, CA, USA) equipped with heated electrospray ionization (HESI) was used for the phenolics analysis. An Agilent 1290 Infinity UHPLC system (Agilent Technologies India Pvt. Ltd., India) was coupled to the mass spectrometer to achieve chromatographic separation. The temperature in the column oven was maintained at 40 °C and in the autosampler was set to 10 °C. Sample was injected using flow-through needle injection mode. To avoid the carryover problem we have enabled needle wash with acetonitrile (0.1% formic acid) before sample injection. Separations were performed using

C-18 column (Shim-pack, ODS III, 2.1 × 150 mm, 2 μm). Mobile phase A was water (10 mM Ammonium acetate) containing 0.1% formic acid, and mobile phase B was acetonitrile containing 0.1% formic acid. Injections of 10 μL were made using flow-through needle mode. A linear gradient was then initiated at a flow rate of 200 μL/min as follows: 5% B at 0–2 min, 25% B at 3 min, 25–100% B at 3–12 min, 100% B at 12–15 min, 5% B at 15.1 min, 5% B at 15.1–22 min. Operating conditions were as follows: spray voltage, 3500 V; ion transfer capillary temperature, 270 °C; source temperature 100 °C; sheath gas 18, auxiliary gas 5 (arbitrary units); collision gas, Argon; S-lens Voltage was optimized for individual metabolites; scan time of 50 ms/transition; and ion polarity positive. The SRM transitions were monitored at unit resolution in quadrupole-1 and quadrupole-3 with a dwell time of 50 ms. They were selected based on the most intense product ion for each metabolite. The typical UHPLC-MS/SRM chromatograms of neem metabolites at the lower and higher concentrations are shown in Fig. 3.2.

The background from the blank injection using methanol showed no interference peaks in any of the analyte channels (Fig. 3.2a). All analytes showed single sharp peak in the C-18 column and eluted in the 20 min gradient. The transitions (703 → 567 m/z) showed peaks in the retention time like Aza A (9.04 min) and another small peak at 9.46 min, might be due to Aza B which normally elutes next to Aza A. Based on the area under the curve, around 6–8% of Aza B is present in Aza A. There are no ISTDs commercially available for all these neem metabolites; we have used estrone-d4 as ISTD for the actual quantification. This shares structural similarity with neem metabolites and elutes in the same way like neem metabolites in the gradient. The ISTD showed almost the same response in both lower and higher concentration of standards and the coefficient of variation is less than 5% in overall sample analysis (Fig. 3.2b, c). In order to quantify the neem metabolites from neem explants, we have constructed the standard curves by taking different concentrations of neem metabolites with same concentration of ISTD and finally plotted ratio (Analyte/ISTD) versus concentration (Kannan et al. 2016).

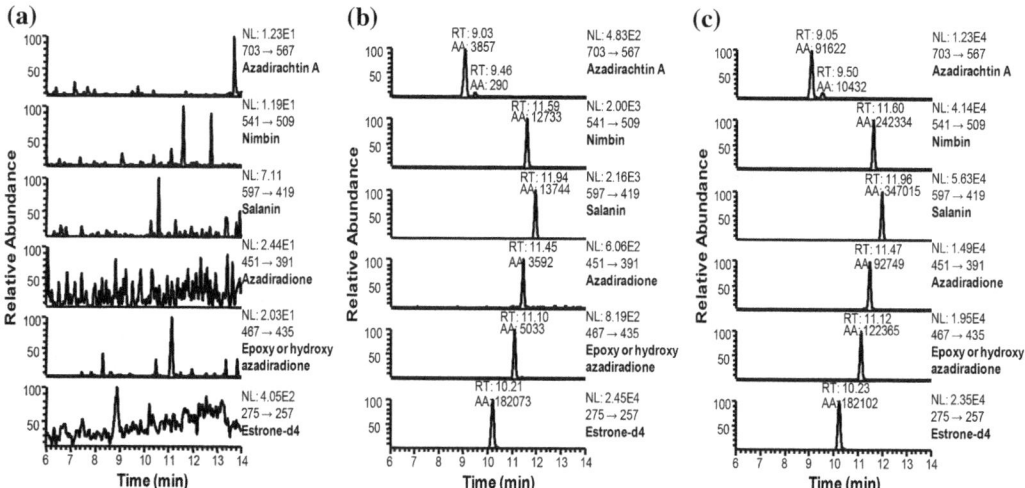

Fig. 3.2 UHPLC-MS/SRM chromatogram of blank, lower and higher concentration of standard neem metabolites. **a** Blank, **b** lower concentration (Aza-312, Nimbin-25, Salanin-125, Azadi-125, and E/H-Azadi—125 pg on column) and **c** higher concentration (Aza-4000, Nimbin-320, Salanin-1600, Azadi-1600, and E/H-Azadi—1600 pg on column) (NL: normalized level)

3.3 Neem Metabolite Extraction from Leaf, Bark and Seed and Analysis by UHPLC-MS/SRM Method

After standardizing the method, we have validated the method by repeating the measurements for five days (inter-day) as discussed in the previous published paper (Kannan et al. 2016). By using the validated method, we have calculated the concentration of neem metabolites from neem explants. We have collected the leaf, bark, and seed from neem tree and allowed to dry at room temperature for three to four days. Dried samples were made into fine powder by using the pestle mortar and stored at 4 °C till the analysis. The metabolites were extracted from 2 mg of each powder by using 1 mL of methanol followed by 5 min sonication in a water bath sonicator and then centrifuged at 10,000 rpm for 5 min. In case of leaf extract, the supernatant was collected after centrifugation (10,000 rpm, 5 min) and 5 µL (10 µg of extract) was diluted to 50 µL and injected 10 µL for the full MS analysis for E/H-azadi. The condition for the analysis is same as standards, instead of SRM scan we have analyzed by full MS scan. The extracted ion chromatogram corresponding to E/H-Azadi (467 Da) and the corresponding mass spectrum are shown in Fig. 3.3. The peak at 435 Da is the product ion of 467 Da, which is happening due to in-source fragmentation of the parent ion. The peak of E/H-Azadi is more in the leaf extract compared to the bark and seed extracts. To analyze in the UHPLC-MS/SRM method, from the supernatant 5 µL was further diluted to 100 µL (20 fold dilution) and 10 µL was further diluted to 50 µL (10 µL sample + 5 µL ISTD + 35 µL methanol) and 10 µL was injected for the analysis. The similar LC-MS conditions used for the standards were also used for the neem explants extracts. All three extracts (leaf, bark, and seed) of neem were quite stable at least for a week in methanol and no degradation was observed at 4 °C. The typical UHPLC-MS/SRM chromatograms of neem metabolites from the neem explants (leaf, bark, and seed) are shown in Fig. 3.4.

All five neem metabolites are present in the seed, whereas azadi in leaf, Aza A and azadi in bark are not detected. In case of azadi channel of leaf extract, there are peaks with the retention time at 10.71 and 11.98 min. Most likely the other peaks corresponding to the isomers of azadi is present in the neem leaf extract. The concentrations of all five neem metabolites were estimated based on our validated method after 100-fold dilution from the original methanol extract. Among the five metabolites analyzed E/H-Azadi is present in very high concentration in neem plant (leaf: 124,239 pg/µg, bark: 906.97 pg/µg, seed: 7309.48 pg/µg) as compared to other metabolites. Interestingly, E/H-Azadi seems to be the most abundant metabolite in the neem leaf and bark extracts and azadi is the highest in the seed extract. In the leaf extract E/H-Azadi is ~136 fold higher compared to bark and ~17 fold higher compared to seed extract. The amount of E/H-Azadi in leaf is 124,239 pg/µg of leaf extract, which constitutes ~10% in the leaf extract.

To further confirm the higher amount of E/H-Azadi in the neem leaf extracts, we have collected neem leaves from 14 different trees in the same day within 5 km^2 and processed in the same way as mentioned above for the analysis. Interestingly, we observed much higher concentration of E/H-Azadi in all 14 independent neem trees (mean: 155,720 pg/µg, (±) SD: 80,430 pg/µg of the leaf extracts). The lowest E/H-Azadi amount estimated is 39,470 and the highest is 309,150 pg/µg from leaf extract. Majority of the leaf extracts ($n = 10$) showed between 100,000 and 300,000 pg/µg of E/H-Azadi. Aza A varies from 0 to 340 pg/µg, nimbin varies from 0.13 to 13.94 pg/µg, Salanin varies from 4 to 95 pg/µg and azadi varies from 0.13 to 75 pg/µg. Based on the mean of all 14 samples the trend in leaf extract looks like E/H-Azadi > Aza A > salanin > Azadi > nimbin. But E/H-Azadi is ~1000 fold higher compared to other metabolites in the leaf extract (Fig. 3.5). Based on the higher amount of E/H-Azadi, we have hypothesized that E/H-Azadi might be one of the important precursors in the actual bio-synthetic pathway of Aza A. Further studies are really

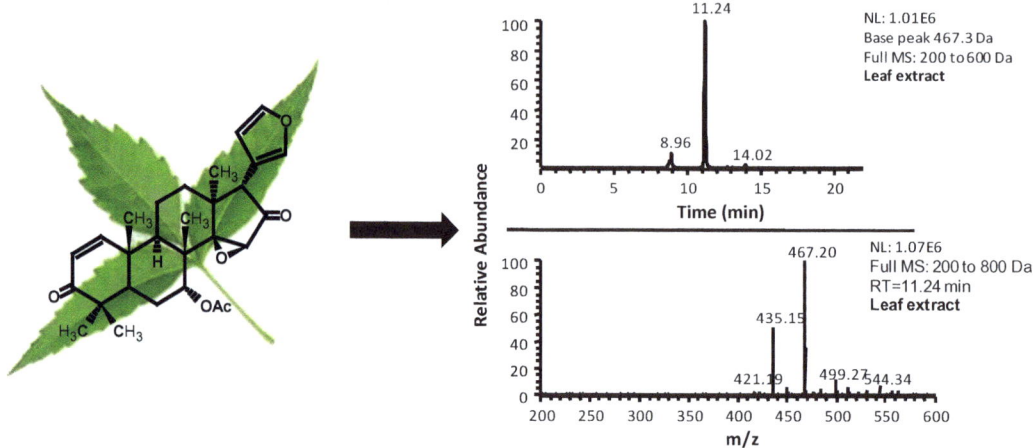

Fig. 3.3 Full UHPLC-MS analysis of E/H-azadiradione in the neem leaf extract

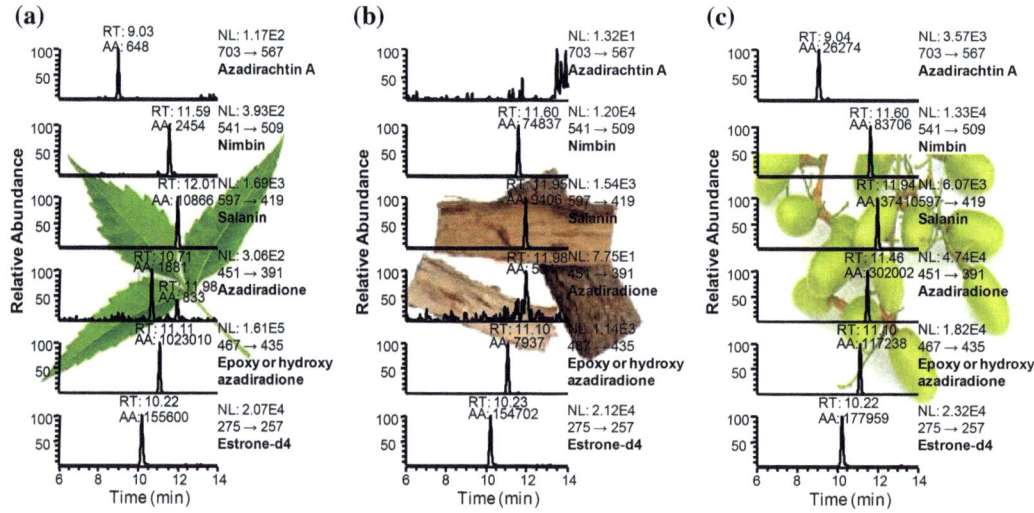

Fig. 3.4 UHPLC-MS/SRM chromatogram of neem explants. **a** Leaf, **b** bark, **c** seed extracts of neem (NL: normalized level)

needed to understand the steps involved in the conversion of E/H-Azadi to Aza A in the biosynthetic pathway. There are many studies showing the anticancer effect using ethanol extract of neem leaf, bark, and seed. Most likely the anticancer property might be due to the presence of E/H-Azadi, which seems to be the most abundant metabolite in the leaf extract. In one of the recent studies, E-Azadi showed anti-inflammatory activity against macrophage migration compared to other four metabolites (Athar et al. 2012).

3.4 Bio-pesticidal Properties of Neem

Contamination of synthetic chemical insecticides in food items and water bodies is becoming one of the major problems in recent years around the

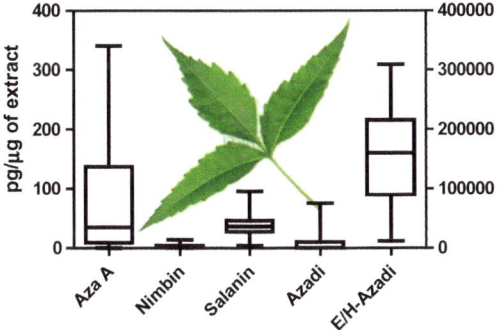

Fig. 3.5 Concentration of neem metabolites quantified from leaf extracts of fourteen different trees. The values presented for Aza-A, Nimbin, Salanin, and Azadi is based the left Y-axis. E/H-Azadi is based on the right Y-axis (∼1000 fold excess compared to other metabolites)

globe. There is an urgent need for effective biodegradable pesticides with greater selectivity. Alternative strategies are to look for the natural biopesticides from medicinal plants like neem to overcome contamination and pest resistance issues (Isman 2006). The botanicals (plant extracts) of neem have enormous potential to influence modern agrochemical research (Schmutterer 1990). Like other plants, neem also has self defence mechanism to protect itself against leaf-chewing insects by producing chemical weapons which are extraordinary. In tests over the last decade, entomologists have found that neem materials can affect more than 200 insect species as well as some mites, nematodes, fungi, bacteria, and even a few viruses (National Research Council (US) Panel on Neem 1992). The neem extract affects members of most, if not all, orders of insects like Orthoptera, Homoptera, Thysanoptera, Coleoptera, Lepidopera, Diptera, Hymenoptera, and Heteroptera (National Research Council (US) Panel on Neem 1992). Among the botanical insecticides currently marketed, neem oil is one of the least toxic to humans and shows very low toxicity to beneficial organisms, so it is, therefore, very promising for the control of many pests (Estefânia et al. 2016). Neem oil contains a group of active ingredients with different chemical characteristics. It was, therefore, believed that the development of insect resistance would be virtually impossible. These features of neem support its contribution to organic agricultural production systems that are more sustainable and do not generate chemical residues (plants and crops are grown without the use of any agrochemicals). This method also helps to maintain soil productivity, ensuring longer production times. Organic agriculture can be a viable alternative production method for farmers, but there are numerous challenges to be overcome. Now there are several companies started selling the neem products as fertilizer and agrochemicals. By replacing chemical pesticides with natural plant-derived biopesticides will have huge positive impact on the global issues like water body contaminations, environmental problems, etc., and definitely will save our future generation by providing green and clean atmosphere.

3.5 Future Perspective

Neem helps to control many of the world's pests and diseases; still many details remain to be fleshed out (National Research Council (US) Panel on Neem 1992). There are challenges in identifying the secondary metabolites responsible for the particular biological action. The metabolomics approach will be of great help in digging the details of the secondary metabolites in the given extracts. Metabolomics defined as "detection and quantification chemical compounds which are less than 1500 Da" from the biological systems. There are challenges in understanding the plant genome and proteome to link with the biological functions of specific medicinal plant extracts. There is a great need for a complementary approach to understand the plant genome and proteome. The only best approach available currently is by looking at the details of plant metabolome either through quantitative metabolomics (UHPLC-MS/SRM methods) or by analyzing via comparative metabolomics (UHPLC-HRMS) by using the advanced liquid chromatography and mass spectrometry (LC-MS) techniques. Unlike genomics and proteomics, analyzing the whole plant secondary metabolome in a single method is cumbersome, due to different chemical categories of compounds. There are compounds under different chemical classes like phenolics, flavonoids, tannins, alkaloids, etc. One can device mass spectrometry based methods for each chemical category to quantify metabolites. In order to develop the quantitative methods, the major hurdle in the field is for the availability standards for the secondary metabolites to develop methods. In case of comparative metabolomics, there is a lack of enough information in the plant metabolome database and also continuous upgradation of softwares to extract information from ion chromatograms. There is a need for regular updates in the existing databases like plant metabolome data base (PMDB), PlantCyc, plant metabolomics.org, etc., which might help us in understanding the details of plant metabolome in the near future.

3.6 Conclusions

We showed here the quantification of five neem metabolites (azadirachtin, nimbin, salanin, azadiradione, and epoxy/hydroxy-azadiradione) from the neem explants like leaf, bark, and seed by using quantitative metabolomics approach. For the first time we have observed here the presence of very high amounts of E/H-Azadi in neem leaf extract, this metabolite might be the common precursor for biosynthesis of Aza. Further studies are needed to understand the role of E/H-Azadi in Aza biosynthesis. The excess amount of E/H-Azadi in the neem leaf might be one of the reasons for its multifunctional properties in nature. Similar kind of methods can also be developed for other metabolites for the plant extracts, which are already been shown to have medicinal properties.

Acknowledgements We would like to thank Department of Biotechnology, India for the financial assistance to setup the Metabolomics Facility at C-CAMP, NCBS Campus.

References

Alexandr YY, Boris VN, Emilie C, Yakov IY (2015) Determination of the chemical composition of tea by chromatographic methods: a review. J Food Res 4(3):56–88

Ali A (1993) Textbook of pharmacognosy. Publication and Information Directorate, New Delhi, India, pp 381–384

Athar A, Saikat H, Hirekodathakallu VT, Rahul K, Manish G, Mohd SI, Chinmay P, Sumanta D, Samik B, Souvik S, Uttam P, Nakul CM, Uday B (2012) Novel anti-inflammatory activity of epoxyazadiradione against macrophage migration inhibitory factor. J Biol Chem 287(29):24844–24861

Avinash P, Devdutta SD, Saikat H, Vairagkar U, Shinde GV, Fayaj AM, Thiagarayaselvam A, Hirekodathakallu VT (2015) Triterpenoid profiling and functional characterization of the initial genes involved in isoprenoid biosynthesis in neem (*Azadirachta indica*). BMC Plant Biol 15(214):1–14

Champagne DE, Koul O, Isman MB, Scudder GGE, Towers GHN (1992) Biological activity of limonoids from the rutales. Phytochemistry 31(2):377–394.

Debas HT, Laxminarayan R, Straus SE (2006) Complementary and alternative medicine, chap 69. In: Jamison DT, Breman JG, Measham AR et al (eds) Disease control priorities in developing countries, 2nd edn. The International Bank for Reconstruction and Development, The World Bank, Washington DC. Available from https://www.ncbi.nlm.nih.gov/books/NBK11796/. Co-published by Oxford University Press, New York

Estefânia VRC, Jhones LDO, Mônica P, Renata DL, Leonardo FF (2016) Neem oil and crop protection: from now to the future. Front Plant Sci 7(1494):1–8

FAO (1997) Medicinal plants for forest conservation and health care. In: Non-wood forest products 11. FAO—Food and Agriculture Organization of the United Nations, Rome, Italy

Fauziah O, Gholamreza M, Sally LTP, Asmah R, Rusliza B, Chong PP (2012) Effect of neem leaf extract (*Azadirachta indica*) on c-Myc oncogene expression in 4T1 breast cancer cells of BALB/c mice. Cell J (Yakhteh) 14(1):53–60

Gordon MC, David JN (2013) Natural products: a continuing source of novel drug leads. Biochim Biophys Acta 1830(6):3670–3695

Hossain MA, Shah MD, Sakari M (2011) Gas chromatography–mass spectrometry analysis of various organic extracts of *Merremia borneensis* from Sabah. Asian Pac J Trop Med 4(8):637–641

Isman MB (2006) Botanical insecticides, deterrents, and repellents in modern agriculture and an increasingly regulated world. Annu Rev Entomol 51:45–66

Kannan R (2014) Sensitive UHPLC-MS/SRM method for quantifying olanzapine metabolites and degradation products from sera. Anal Methods 6:5250–5257

Kannan R, Varalaxmi BA, Malali G (2016) UHPLC-MS/SRM method for quantification of neem metabolites from leaf extracts of Meliaceae family plants. Anal Methods 8:2020–2031

Kaushik SH, Lakshmi M, Rajendra H, Anil K (2014) NeeMDB: convenient database for neem secondary metabolites. Bioinformation 10(5):314–315

Kokate C, Purohit AP, Gokhale SB (2010) Pharmacognosy. Nirali Prakashan, Maharashtra, India

Mamta S, Jyoti S, Rajeev N, Dharmendra S, Abhishek G (2013) Phytochemistry of medicinal plants. J Pharmacogn Phytochem 1(6):168–182

Mohammad AA (2016) Therapeutics role of *Azadirachta indica* (Neem) and their active constituents in diseases prevention and treatment. Evid Based Complement Altern Med 2016(Article ID 7382506):11. http://dx.doi.org/10.1155/2016/7382506

Nagesh AK, Vijay Y, Kannan R, Mahesh HB, Anantharamanan R, Meghana S, Heikham R, Ramya ML, Chandana SL, Shilpa S, Aishwarya R, Sathyanarayana BN, Malali G (2015) Comprehensive analyses of genomes, transcriptomes and metabolites of neem tree. Peer J 3:e1066

National Research Council (1992) Neem: a tree for solving global problems. The National Academies Press, Washington DC. https://doi.org/10.17226/1924

National Research Council (US) Panel on Neem (1992) Effects on insects, chap 5. In: Neem: a tree for solving global problems. National Academies Press (US), Washington DC. Available from https://www.ncbi.nlm.nih.gov/books/NBK234642/

Nivedita N, Padma R, Vairavan L, Dasaradhi P, Kannan R (2015) A quantitative metabolomics peek into planarian regeneration. Analyst 140:3445–3464

Priyadarshini K, Keerthi AU (2012) Paclitaxel against cancer: a short review. Med Chem 2(7):139–141

Ruchi T, Amit KV, Sandip C, Kuldeep D, Shoor VS (2013) Neem (*Azadirachta indica*) and its potential for safeguarding health of animals and humans: a review. J Biol Sci. https://doi.org/10.3923/jbs.2013

Saikat H, Fayaj AM, Thiagarayaselvam A, Devdutta SD, Hirekodathakallu VT (2014) Expedient preparative isolation and tandem mass spectrometric characterization of C-seco triterpenoids from neem oil. J Chromatogr A 1366:1–14

Schmutterer H (1990) Properties and potential of natural pesticides from the neem tree, *Azadirachta indica*. Annu Rev Entomol 35:271–297

Selma B, Eltahir AK, Sami AK, Mohammad AA, Abdlmarouf M, Yousef HA, Paul FSE, Mohammed F (2013) *Azadirachta indica* ethanolic extract protects neurons from apoptosis and mitigates brain swelling in experimental cerebral malaria. Malaria J 12(298):1–9

Simon JH, Benjamin PYL, Rainer S, Rudolf K (2014) Liquid chromatography-mass spectrometry for the determination of chemical contaminants in food. Trends Anal Chem 59:59–72

Tan QG, Luo XD (2011) Meliaceous limonoids: chemistry and biological activities

Vincent PKT, Denis Z, Moses NN (2008) The antimalarial potential of medicinal plants used for the treatment of malaria in cameroonian folk medicine. Afr J Tradit CAM 5(3):302–321

Wungsem R, Sreya D, Debajyoti D, Jayram H (2013) A brief review on the botanical aspects and therapeutic potentials of important Indian medicinal plants. Int J Herb Med 1(3):38–45

Utilization of Neem and Neem Products in Agriculture

4

Rishu Sharma and Chittaranjan Kole

Abstract

The excessive dependence on pesticides and fungicides to protect the crops from various pathogens and pests has caused irreparable losses to our land and water resources in addition to severe degradation of soil quality and beneficial soil micro-flora. In this chapter, we have focused on the use of neem and its products in agriculture to curb and mitigate the losses caused by chemicals which have been used indiscriminately and incessantly since past decades. Synthetic pesticides enable safer control of pest population but these neem-based products come with few challenges as well like short shelf life, photosensitivity, and volatilization which causes hindrance in their application at commercial level.

4.1 Introduction

Neem has been mainly cultivated in Indian sub-continent especially under arid climates. The utilization of Neem and its products has been an indispensable part of agriculture since ages. Azadirachta indica is the scientific name of locally known Neem tree. It belongs to the family Meliaceae which has originated from Indian sub-continent (Campos et al. 2016). The genus Azadirachta consists of one or two species within it. It is native to Asian sub-continent viz. India, Bangladesh, Thailand, Nepal, and Pakistan. It has the potential to grow well in tropical and subtropical regions. In India, it is called "Divine Tree", "Life-giving tree", "Nature's Drugstore", "Village Pharmacy", and "Panacea for all diseases" (Hossain and Nagooru 2011). In the present scenario, Neem has been identified as an efficient natural product which has enormous applications in worldwide agriculture in combating global agricultural and environmental issues (Biswas et al. 2002; Subapriya and Nagini 2005). The transition from use of synthetic products to natural ones is evident in agricultural industry to prevent or reduce the losses caused by insects, pests, and pathogens to improve the yield ad quality of the produce (Oerke and Dehne 2004; Cooper and Dobson 2007). The excessive use of synthetic insecticides has resulted in a number of serious problems like development of insect resistance to insecticides, harm to other natural enemies of insects, toxic effects on plants and soil, etc. (Chandler et al. 2011; Damalas and Eleftherohorinos 2011). The synthesis of the chlorinated pesticide dichlorodiphenyltrichloroethane (DDT) and organophosphorus pesticides, research on biological control methods

R. Sharma (✉)
Department of Plant Pathology, BC Agricultural University, Mohanpur, West Bengal 741252, India
e-mail: rrishu.sharma90@gmail.com

C. Kole
National Research Centre on Plant Biotechnology, New Delhi 1100012, India

reduced imperatively (Doutt 1964; Niu et al. 2014). In agriculture, neem can be utilized efficiently by using its products like Neem oil, fruit and by-products like seed cake are used as biopesticides, fungicides, and organic manures. In the traditional Indian farming system, pest management, nutrient supply were integrated into the cropping system using locally available plant-derived substances like Neem. These natural extracts protected the crops and nourished the soils. It is evident that the traditional Indian practices were very sound and are still in use. Today with advanced technologies, neem extracts are further processed to produce a wide range of products as mentioned above, suitable to be used in agriculture. Incorporating latest scientific and technical developments, neem products are suited to agricultural growth. Extensive toxicological studies have been carried out on Neem products worldwide and have been finally cleared for use in horticulture by the Environmental Protection Agency (Anonymous 2009).

Neem products are proven effective against thrips, whiteflies, aphids, leaf miners, bugs, and a large number of other key insect pests. They act as natural insect repellants. Azadirachtin and other bitter compounds found in Neem make insects stay away from treated plants. It has also been studied that the amount of Azadiractin in the seeds can be increased by artificial infection with arbuscular mycorrihzea (Venkateshwarlu et al. 2008). They are like synthetic pesticides. Neem can reduce overall insecticide, pesticides, fumigants, etc. can act as various other replacements for sustainable and eco-friendly cultivation practices and environment. Another important advantage of neem products in agriculture is that they are safe for the workers (Campos et al. 2016). There are no handling risks and can be used throughout the entire crop production cycle.

4.2 Importance of Neem in Plant Nutrition

Neem and its by-products are a major source of nutrients for plants. In traditional farming system employed neem extracts for pest management and to supply nutrients to plants (Mossini and Kemmelmeier 2005; Sujarwo et al. 2016). Neem and its products can be used in different formulations and amendments as below:

4.2.1 Neem Seed Cake

Neem cake is the residue left after oil has been pressed from neem seeds. It is used as an organic manure resulting in high yield of crops and plants. The chemical composition of Neem Seed Cake consists of Azadirachtin, Nitrogen, Phosphorus, Potassium, Carbon, Sulphur, Calcium, Magnesium. It comes in different formulations.

4.2.1.1 Neem Cake Granules

Neem cake granules are used as a natural and environmental friendly fertilizer and manure in farming and agriculture. Neem cake is the residue left after oil has been extracted from neem seeds or kernel. It has the highest azadirachtin content as compared to other parts of the neem tree. Chemical Composition of Neem Cake Granules, composes 3.56% of Nitrogen, 0.83% Phosphorus, 1.67% Potassium, 0.99% calcium and 0.75% Magnesium.

Advantages These are 100% natural and without any side effects, less quantity is required as compared to synthetic fertilizers, nontoxic, Pest repellent (Table 4.1).

4.2.1.2 Neem Cake Powder

Neem cake powder is being widely used in organic farming and agriculture. This natural product is rich in nitrogen, phosphorus, sulphur, and calcium, required for growth and high yield of crops. When mixed with soil, it increases the nutrient content and fertility. The chemical Composition of Neem Cake Powder: Nitrogen, 3.56%, Phosphorus, 0.83%, Potassium, 1.67%, Calcium, 0.99%, Magnesium, 0.75%.

Uses of Neem Cake Powder It works as a soil conditioner, nitrogen saver, fertilizer, manufacture of pesticides and insecticides, used in production of various food and cash crops.

Table 4.1 Recommendations of use of neem oil and NSKE for different pest management in paddy crop

Pest	Crop	Formulation
Brown plant hopper *Nilaparvata lugens*	Rice	Neem oil 3%, neem seed kernel extract 5% or soil application @ 25 kg/ha
Black bug *Cotinophoralurida*	Rice	Spray neem seed kernel extract 5% or soil application @ 25 kg/ha
Earhead bug *Leptocorisaacuta*	Rice	Spray neem seed kernel extract 5% or soil application @ 25 kg/ha
Sheath rot *Saroclodium oryzae*	Rice	NSKE (5%) or neem oil 3%
Sheath Blight *Rhizoctonia solani*	Rice	Foliar spray with neem oil at 3%
Bacterial leaf blight *Xanthomonas oryzae pv. oryzae*		Spray neem oil 3% or NSKE 5%

4.2.1.3 Neem Cake Manure

Neem cake has been traditionally used as an organic or natural manure and according to recent studies undertaken, it helps in a greater crop yield as compared to the synthetic manures. It also acts as a fertilizer and pest repellent at the same time. It helps to increase the productivity and fertility of soil. Neem cake powder and neem cake granules are used to manufacture this eco-friendly manure without any after-effects. It can also act as a coating agent when combined with urea. Neem cake manure also helps soil retain the nitrogen quantity. It does not have any negative effects on other living organisms.

Benefits of Neem Cake Manure Natural and free from mud, stones, etc., less quantity required as compared to synthetic manures, increases the nitrogen content of soil thus increasing the crop yield in the long run, environmental friendly, does not have any negative effects on other microbial organisms, makes the soil more fertile, and better than farm manure.

4.2.1.4 Neem Cake Bio Mix

It is a kind of bio and organic fertilizer with neem and its parts one of the main constituents. It is used in the cultivation and growth of plants and crops. One of the most essential agricultural inputs, it is widely used to enrich the soil by providing required nutrients.

4.2.1.5 Neem Cake Fertilizer

Neem cake is the richest and probably the best source of manufacturing organic fertilizer. It is the most favored agricultural product used for various crop management. It is available in granular and pellet form, it helps to enrich the soil and aids in faster growth of plants. It acts as a biofertilizer and helps in providing the required nutrients like nitrogen, phosphorus to plants. It is widely used to ensure a high yield of crops. Neem is used as a fertilizer both for food as well as cash crops, particularly vegetables and sugarcane. Neem Cake Fertilizer is an excellent organic fertilizer. Its essential contents and other micronutrients are Nitrogen, Phosphorous, Magnesium, Potassium, Calcium, Sulphur, Zinc, Copper, Iron, and Manganese.

Benefits of Neem Cake Fertilizer It performs the dual function of both fertilizer as well as pesticide, acts as a soil enricher, reduces the growth of soil pest and bacteria, provides macronutrients essential for all plant growth, helps to increase the yield of plants in the long run, biodegradable and eco-friendly, and excellent soil conditioner.

4.2.2 Neem Oil

Neem oil is an essential vegetable oil extracted from the seeds and fruits of neem tree. Neem oil contains various compounds that have medicinal, cosmetic and insecticidal properties. In fact, it is the most important and useful by-products of the magical tree (Koul et al. 2004). Though the neem trees are commonly found throughout Asia and India and neem products are used to treat a variety of medical conditions in Asian countries, nowadays the usage of neem products is seen in western countries as well.

4.2.2.1 Neem Oil Characteristics

Neem oil is dark to light brown in color with very strong smell. Bitter smell and taste of the neem oil is because of triglycerides and triterpenoid compounds present in the oil. Neem oil also contain omega-6, omega-9, palmitic acid, stearic acid, omega-3, and palmitoleic acid.

4.2.2.2 Neem Oil Applications in Agriculture

There is an ingredient in neem oil, azadirachtin, which is naturally found in neem seed oil is responsible for acting as natural Pesticide, natural Insecticide, natural Fungicide. Neem oil has been used for hundreds of years in controlling plant pests and diseases. Many studies have shown that the spray solution of neem oil helps to control common pests like whiteflies, scales, spider mites, locusts, mealy bugs, Japanese beetles, etc. (Table 4.1).

4.2.2.3 Neem Oil Processing

Mechanical Press Method This method is one of the oldest methods of processing oil. Seeds are placed in a tub or container and a form of press or screw is used to squeeze the seeds until the oil is pressed out and collected.

Steam and High-Pressure Method This method makes use of high-pressure extraction method to squeeze out oil from seeds. Seeds are heated in steam and under high pressure enabling maximum extraction of oil. This method is not very good as most of the active ingredients and compounds are destroyed by high temperature.

Solvent Extraction Method One of the most used methods of extracting neem oil, it uses a solvent, preferably a petroleum solvent/alcohol solvent for processing oil. It ensures maximum extraction of oil.

Cold Pressed Method This method of extracting oil is the most used by leading manufacturers though it is more expensive than the other methods.

4.2.2.4 Types of Neem Oil

Neem Essential Oil
Neem Seed Oil

(a) **Neem Essential Oil**

It is made from plant extracts and they have been in use since long time because of their therapeutic value. They are used to manufacture natural or herbal cosmetics, herbal medicines and are being increasingly used in aromatherapy. They are known to have a healing and calming effect on human mind body and soul. They act as skin soothers and rejuvenators. Essential oils also help to keep the hair and skin clean by detoxifying the body. Neem is being used for its therapeutic and medicinal benefits to manufacture essential oils. They are being manufactured and exported on a commercial basis for use in home remedies and aromatherapy.

Properties of Neem Essential Oil

(a) Antifungal
(b) Antibacterial

(b) **Neem Seed Oil**

Neem seeds are processed to get neem oil which can be put to a number of uses like manufacturing natural or biopesticides, insecticides, fungicides, etc. Commercial production of neem tree for its seeds and other parts is quite a few countries like India, Burma, etc. (Anonymous 1997; Marz 1989). Neem seed oil is very bitter with garlic-like smell, it is however quite rich in fatty acids and other essential amino acids.

Percentage of Fatty acids found in Neem Seed Oil

Linoleic acid	2%
Oleic acid	50%
Palmitic acid	12%
Stearic acid	20%
Other lower fatty acids	2%

Percentages may vary in samples and depending on the place and time of collection of seeds.

Use of Neem Seed Oil in Agriculture The principle ingredient in Azadirachtin found naturally in neem seed oil is being used the world which appears to cause 90% of the effect on most pests over for manufacturing, natural pesticide, natural insecticide, natural fungicide (Fig. 4.1). Azadirachtin A is the most active biological ingredient which shows insecticidal property compared to other analogs of azadirachtin (Sola et al. 2014). Neem oil is mainly extracted from the seeds but some other parts too are used for oil extraction despite having lower percentage of Azadirachtin (Nicoletti et al. 2012).

4.2.3 Neem Leaf Extract

Neem leaf extract has a fruit-like smell and contains essential fatty acids; this extract finds large-scale personal and industrial application. In agriculture, they are used in a number of pesticides and insecticides. In a recent study, neem leaf extract was used in studying the biofilm inhibition formation by *Pseudomonas aeruginosa* (Harjai et al. 2013). Factors contributing to adherence and biofilm formation were observed and demonstrated that neem leaves extract was quite effective in disrupting formation and structure of biofilms.

4.2.3.1 Process of Neem Leaf Extraction

Mature green neem leaves are collected and allowed to dry partially. These dried neem leaves are then crushed and powdered. The crushed neem leaves are then subjected to either aqueous or organic solvent to get a concentrated extract. For making neem leaf extract, certain extraction process utilizes carbon dioxide at critical temperatures and pressures to extract the active ingredients of the neem leaf, the usual high temperatures or harsh chemicals are done away with, resulting in a better concentrated and potent extract.

Use in Agriculture Used to manufacture natural and organic pesticides and neem insecticides, can be used as an antifeedant and helps in the growth and promotion of plants.

4.3 Neem and Its By-Products in Plant Protection

4.3.1 Biopesticide

The importance of neem as biopesticide was realized by the modern scientific community, as early as 1959, when a German scientist in Sudan found that neem was the only tree that remained green during a desert locust plague (Schmutterer 1990; Rajendran 2010) Literatures confirm that neem can effectively get rid of over 200 pest species that affects plants (Sarup and Srivastva 1971; Wilps et al. 1992). A major crops are affected by a number of pests which are responsible for decreasing the average yield of the crop produce (Table 4.2). The pesticidal characteristics of neem is largely attributable to Azadirachtin found in the neem extracts which is a growth regulator and as well as a powerful feeding and ovipositional deterrent (Bramhachari 2004). Azadirachtin is nonvolatile and an insect cannot prevent it by smell but has to taste it, in order to respond to it. A taste of azadirachtin stimulates at least one 'deterrent neuron' in insects which show an antifeedant response (Schmutterer 1990). The strength of 'deterrent

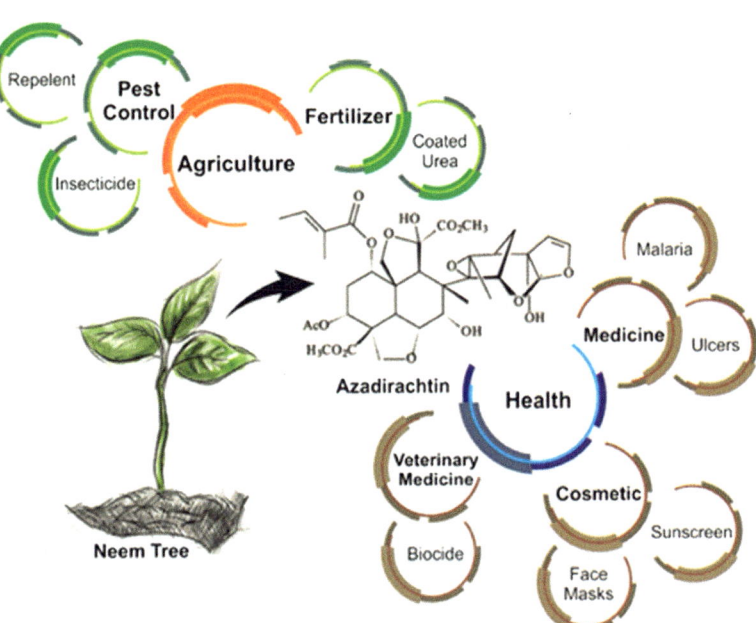

Fig. 4.1 Potential applications of Azadirachtin in different areas. *Source* Campos et al. (2016)

Table 4.2 Major Indian agricultural crops/commodities and its associated major pests

Sr. no.	Crops/commodities	Major pests
1.	Rice	*Acrida exaltata, Ampittia dioscorides, Chilo polychrysa, Cnaphalocrocis trapezalis, Lenodora vittata, Leptocorisa acuta, Nephotettix parvus*, and many more
2.	Maize	*Agonoscelis nubilis, Aloa albistriga, Anomala dimidiate, Mythimna loreyi, Myllocerus viridanus*, and many more
3.	Sugarcane	*Abdastartus atrus, Chilo sacchariphagus indicus, Cofana spectra, Mocis frugalis, Mythimna separate*, and many more
4.	Groundnut	*Agrius convolvuli, Atractomorpha crenulata, Caryedon serratus, Dudua aprobola, Sphenoptera perroteti, Spodoptera litura*, and many more
5.	Cotton	*Aloa albistriga, Anomis flava, Bemisia tabaci, Ferrisia virgata, Helicoverpa armigera, Phenacoccus solenopsis, Plautia crossota*, and many more
6.	Banana	*Aleurodicus rugioperculatus, Aularches miliaris, Bactrocera dorsalis, Cosmopolites sordidus, Hishimonus phycitis, Odoiporus longicollis, Prodromus clypeatus, Parasa lepida*, and many more

neuron' responses has been correlated with the strength of antifeedant responses. Neem oil can also suffocate mites, whiteflies, aphids and other types of soft-bodied insects on contact. So it is clear that neem does not kill on contact, rather it inhibits feeding and reproduction of the pests. These multiple modes of action make it unlikely that insects and plant pathogens can develop resistance to nee. Also, certain pest such as floral thrips, diamond back moth and several leaf miners which develop resistance to the inorganic pesticides or that are inherently difficult to control with conventional pesticides are effectively controlled or managed with neem. Neem-based products can act as antifeedants, growth regulators, sterilants, anti-oviposition agents, and repellents due to their compositional complexity (Gonzalez-Coloma et al. 2013).

Neem pesticides play a vital role in pest management and hence have been widely used in agriculture. There has been an evident shift all over the world from synthetic pesticides to non-synthetic ones; this is largely because of the widespread awareness of the side effects of these synthetic pesticides not only on plants and soil but also on other living organisms. This is a great opportunity for neem pesticides manufacturers to cash in on the growing popularity of natural or herbal pesticides. Neem pesticides are being manufactured and exported to various countries as a lot of research has been conducted to test the safety and efficacy of neem for use as a pesticide (Anis Joseph et al. 2010; Vethanayagam and Rajendran 2010). Azadirachtin is the main ingredient used to manufacture biopesticides. Neem oil and seed extracts are known to possess germicidal and antibacterial properties which are useful to protect the plants from different kinds of pests. One of the most important advantages of neem-based pesticides and neem insecticides is that they do not leave any residue on the plants.

4.3.2 Bio-Fungicide

As a fungicide, neem oil is mainly used as a preventative and when disease is just starting to show. It coats the leaf surface which in turn prevents the germination of the fungal spores. Neem oil is effective against rots, mildews, rusts, scab, leaf spot, and blights. *Fusarium oxysporum* causes Fusarium wilt disease in different plants (Shanmugan et al. 2007; Ploetz 2000). *Rhizoctonia solani* and *F. solani* causes damping off (Sturrock et al. 2015) while *Sclerotinia sclerotiorum* causes stem rot (Mueller et al. 2002; Young et al. 2004). Leaf extract of neem can inhibit the aflatoxin production as well as *Aspergillus parasiticus* growth (Ghorbanian et al. 2007; Allameh et al. 2002). Antifungal effects of neem leaf extract also reported from south America against *Crinipellis perniciosa* and *Phytophthora* species causing Witches broom and Pot Not of cocoa (Ramos et al. 2007). The antifungal effects of neem leaves and seed extracts obtained by ethanol, hexane, and petroleum ether were examined separately in vitro against *F. oxysporum*, *R. solani*, *Alternaria solani,* and *S. sclerotiorum* indicated that seeds and leaves extracts could cause growth inhibition of tested fungi, all these extracts and concentrations of extract inhibited the growth of pathogenic fungi at a significant level. The azadirachtin, nimonol, and expoxyazdirodione were detected from neem extract by using High-Performance Liquid Chromatography (HPLC) and concluded that neem leaf and seed extracts were effective as antifungal against all tested fungi but *F. oxysporum* and *R. solani* were the most sensitive fungi (Moslem and El-Kholie 2009).

4.3.3 Bioinsecticide

The transition from use of synthetic products to natural ones is evident in agricultural industry also; Excessive use of synthetic insecticides has resulted in a number of serious problems like development of insect resistance to insecticides, harm to other natural enemies of insects, toxic effects on plants and soil, etc.

4.3.3.1 Why Neem Bioinsecticides are Preferred Over Other Insecticides?

Neem is being used to manufacture what is known as the natural or bioinsecticide, that are environmental friendly and do not have any toxic effects on plants and soil. Neem insecticide are used to protect both food as well as cash crops like rice, pulses, cotton, oils seeds, etc. It is vital for all crops, trees, plants, flowers, fruits, veggies around the home as well as organic and commercial growers. Active ingredient Azadirachtin, found in neem tree, acts as an insect repellent and insect feeding inhibitor, thereby protecting the plants. This ingredient belongs to an organic molecule class called tetranortriterpenoids. It is similar in structure to insect hormones called "ecdysones," which control the process of metamorphosis as the insects pass from larva to pupa to adult. It is interesting to note, that neem doesn't kill insects, but alters their life process. There has recently been increased interest in the application of plant-based materials (botanical insecticides), such as neem oil, in pest control. These products have been found to be safer for the management of pests as compared to application of synthetic chemicals. The neem-based materials are vital in Integrated Pest Management. Regular scouting helps to insure the early detection of immature stages of many insects. Azadirachtin is compatible with insecticidal soap, superior horticultural oil and *Bacillus thuringiensis* (*Bt*). Several studies have investigated the relationships between botanical insecticides and natural enemies of agricultural pests (Islam et al. 2011; Mamoon-ur-Rashid et al. 2011; Islam and Omar 2012; Tunca et al. 2012) evaluated the use of different neem-based products in colonies of *Beauveria bassiana, Isaria fumosoroseus*, and *Lecanicillium lecanii*, and the results showed that these entomopathogenic fungi were compatible with most products tested. Raguraman and Kannan (2014) conducted a review in order to score the impact and safety of different botanical insecticides in the presence of parasitoids and predators (beneficial arthropods), with the aim of standardizing strategies and application methods to achieve better management of agricultural pests.

4.3.3.2 Parts of Neem Seed to Manufacture Insecticides

(a) **Neem Oil Insecticide**: The major parts/extracts of neem seed that are used for making neem insecticides are the neem seed kernels and the neem seed oil. The oil is considered a contact insecticide with a broad spectrum of action (Cox 2002).

(b) **Neem Seed Kernels**: According to recent studies conducted on parts of neem, it was found that neem seed extracts contain azadirachtin, which in turn works by inhibiting the development of immature insects (Table 4.3).

(c) **Neem Oil**: Neem oil or the neem seed oil is extensively used to manufacture insecticides used for different crops. Neem oil enters the system of the pests and obstructs their proper working. Insects do not eat, mate, lay eggs resulting in the breaking of their life cycle. Another interesting function of neem oil pesticides is that they do not harm the beneficial insects. The neem oil insecticides only target the chewing and sucking insects.

Table 4.3 Specification for neem kernel oil

Characteristic	Requirement
Maximum moisture and insoluble impurities lovibond color (1/4 in cell), expressed as y+ 5R, not deeper than	0.3% by weight 45.0
Refractive index at 40 °C	1.4615–1.4705
Specific gravity	0.908–0.934
Saponification value	180–205
Iodine value (Wij's method)	65–80
Maximum acid value	15
Maximum unsaponifiable matter	2% by weight
Minimum titre	36 °C

Source Indian Standards Institution specification 4765. The color index indicates the degree of deterioration of nonfatty constituents present in the seed vis-à-vis quality of oil

4.3.4 Bio-Fumigant

Neem tree has been used against household, storage pests and crop pests. Neem pest fumigant is available in gaseous state and is used as a pesticide and disinfectant. It is being used by a large number of countries on a commercial basis by farmers and agriculturists. This 100% natural product is being exported as it is nontoxic and does not affect the environment. It assumes more importance in developing countries where millions of deaths are reported every year due to the accidental intake of synthetic pest fumigants. This natural fumigant not only kills pests but also affects them negatively by acting as feeding and oviposition deterrence, mating disruption, inhibition of growth, etc. According to studies undertaken, neem fumigant helps to protect stored rice grains from pests. One of the major benefits of this organic fumigant is that pests do not develop resistance to it. With the increasing trend of using biofertilizers, insecticides and pesticides, neem is being increasingly cultivated and grown all over the world to get active ingredient azadirachtin, responsible for stopping the growth cycle of insects and pests, fungi, etc. Neem is also assuming a lot of importance in crop management. Considering the fact that neem is not only a cheaper, naturally occurring product and an effective method to control pests and insects, but also has no side effects on plants or other living beings, it is not a wonder that researches are being carried to try neem and its products for large-scale production of natural pesticides and insecticides. This is a good opportunity for manufacturers and exporters to produce quality bio agricultural products. Neem oil and seed extracts are known to possess germicidal and antibacterial properties which are useful to protect the plants from different kinds of pests. This natural product does not leave any residue on plants.

4.3.4.1 Benefits
Neem fumigants are eco-friendly, do not harm other microorganisms, are nontoxic, and do not contaminate terrestrial and aquatic environment. Pests do not develop resistance to it, there are no negative after effects, are relatively less expensive, are pest repellent and nourish the soil and function as pest reproduction controller.

4.3.4.2 Parts of Neem Used for Manufacturing Pest Fumigant
Neem oil and seed extracts are known to possess germicidal and antibacterial properties which are useful to protect the plants from different kinds of pests. This natural product does not leave any residue on plants.

4.3.4.3 Benefits of Neem Pest Fumigant
Eco-friendly, does not harm other microorganisms, nontoxic, does not contaminate terrestrial and aquatic environment, pests do not develop resistance to it, no negative after effects, relatively less expensive, pest repellent, nourishes the soil, pest reproduction controller.

4.3.5 Biofertilizer

The material left after oil is squeezed out from seeds and is popularly known as the seed cake; it acts as a biofertilizer and helps in providing the required nutrients to plants. It is widely used to ensure a high yield of crops. Neem is used as a fertilizer both for food crops and cash crops, particularly rice and sugarcane crop.

4.3.5.1 Neem Fertilizer
Fertilizers are organic or inorganic plant foods which are available either in liquid or granular form and are used to amend the soil in order to improve the quality or quantity of plant growth. With the growing dissatisfaction with synthetic fertilizers, in various countries, bio and natural fertilizers are gaining popularity. Neem fertilizer is also in great demand as they are not only eco-friendly, but also have no after effects. It has shown great potential for exports in the overseas market.

Table 4.4 Requirements of neem cake for manuring

Characteristic	Requirement
Maximum moisture (% by mass)	10.0
Minimum water-soluble organic N% by mass on moisture-free basis	2.5
Maximum total ash (% by mass)	13.0
Maximum acid insoluble ash (% by mass) on moisture-free basis	5.0

Source Indian Standards Institution specification 8558

4.3.5.2 Parts of Neem Used as Fertilizer

Neem Seed Cake Matter left after oil is squeezed out from seeds is popularly known as the seed cake; It acts as a biofertilizer and helps in providing the required nutrients to plants. It is widely used to ensure a high yield of crops. Neem is used as a fertilizer both for food as well as cash crops, particularly vegetables and sugarcane.

4.3.5.3 Benefits of Neem Fertilizer

- Performs the dual function of both fertilizer as well as pesticide
- Acts as a soil enricher
- Reduces the growth of soil pest and bacteria
- Provides macronutrients essential for all plant growth
- Helps to increase the yield of plants in the long run
- Biodegradable and Eco-friendly
- Excellent soil conditioner.

4.3.6 Bio-Manure

Manure is any animal or plant material used to fertilize land especially animal excreta for improving the soil fertility and thus promoting plant growth. Neem manure is gaining popularity because it is environmental friendly and also the compounds found in it help to increase the nitrogen and phosphorous content in the soil. It is rich in sulphur, potassium, calcium, nitrogen, etc. Neem cake is used to manufacture high quality organic or natural manure, which does not have any aftermaths on plants, soil, and other living organisms. It can be obtained by using high technology extraction methods like cold pressing or other solvent extraction. It can be used directly by mixing with the soil or it can be blended with urea and other organic manure like farmyard manure and seaweed for best results. Benefits: It is biodegradable and eco-friendly, nourishes the soil and plants by providing all the macro and micronutrients, helps to eliminate bacteria responsible for denitrifying the soil, ideal for cash crops and food crops, increases the yield of crops, helps to reduce the usage of fertilizer, thus reducing the cost of growing plants, antifeedant properties that help to reduce the number and growth of insects and pests (Table 4.4).

4.3.6.1 Part of Neem Used to Manufacture Manure

Neem cake is used to manufacture high quality organic or natural manure, which does not have any aftermaths on plants, soil, and other living organisms. It can be obtained by using high technology extraction methods like cold pressing or other solvent extraction.

4.3.6.2 Usage of Neem Manure

It can be used directly by mixing with the soil or it can be blended with urea and other organic manure like farmyard manure and seaweed for best results.

4.3.6.3 Benefits of Neem Manure

- Biodegradable and Eco-friendly.
- Nourishes the soil and plants by providing all the macro and micronutrients.
- Helps to eliminate bacteria responsible for denitrifying the soil.
- Ideal for cash crops and food crops.
- Increases the yield of crops.

- Helps to reduce the usage of fertilizer, thus reducing the cost of growing plants.
- Antifeedant properties help to reduce the number and growth of insects and pests.

4.3.7 Neem Compost

Compost is a mixture of decaying organic matter, as from leaves and manure and used in improving the soil quality and fertility; also providing essential nutrients to plants and soil. Neem compost is being used by farmers and agriculturists on a large scale because of its benefits. It helps converting the organic matter into humus. Better than other manures, as they do not have any after-effects. It is being manufactured and exported in various countries because of its effectiveness and use for almost all food and cash crops. It is generally free from materials like mud and dirt. Neem leaves are used to manufacture organic and natural compost.

4.3.7.1 Benefits of Neem Compost

Increased plants resistance to pest attacks, environmental friendly, universal application for all crops and plants, rich in vitamin, minerals and enzymes, increases the nitrogen percentage in soil, retains freshness of plants and vegetables for longer duration, best soil conditioner, nontoxic, protects the plants from certain root diseases, high nematicides property.

4.3.8 Urea Coating Agent

Neem and its parts are being used to manufacture urea coating agent to improve and maintain the fertility of soil. Nitrogen is one of the main nutrients required by plants for their development, and it is predominantly supplied by urea (Ni et al. 2014). The fertility of the soil can be measured by the amount of nitrogen, potassium, and phosphorous it has; there are certain bacteria found in soil, which denitrify it. Use of neem urea coating agent helps to retard the activity and growth of the bacteria responsible for denitrification. It prevents the loss of urea in the soil.

In order to avoid the nitrogen losses, the control of urea hydrolysis and nitrification is one of the principal strategies. The bacterial activity that is responsible for denitrification and thus decreasing the loss of urea from the soil, neem act as nitrification inhibitor (Musalia et al. 2000; Mohanty et al. 2008). It can also be used to control a large number of pests such as caterpillars, beetles, leafhoppers, borer, mites, etc. Urea coating is generally available either in liquid form or powdered form.

4.3.8.1 Properties of Neem Urea Coating Agent

- Antifeedant
- Antifertility
- Pest growth regulator.

4.3.8.2 Benefits of Neem Urea Coating Agent

Excellent soil conditioner, natural or biopesticides, environmental friendly, nontoxic, reduces urea consumption, convenient and easy to apply, high soil fertility, and increases the yield of crops.

4.3.9 Soil Conditioners

Neem is a natural soil conditioner that helps improve the quality of soil, thereby enhancing the growth of plants and fruits. It not only helps the plants grow but also prevents them from being destroyed by certain pests and insects. Organic soil conditioner is gaining popularity in agricultural industry, not only in Asian countries like India but also the western counterparts such as the USA, the UK, and Australia. This is so, because they are organic, have no harmful effects, and cheaper than the other soil conditioners. This natural soil conditioner is also multifunctional. Coffee planters in the subtropical regions are also using neem soil conditioner for good coffee plantation. The planters are aware of its soil enhancing properties which is responsible for increasing the soil fertility. Neem seed granules or powdered seeds are used to manufacture the soil conditioner; it can be applied during sowing of plants or can be sprinkled and raked into the soil. The process of sprinkling should be followed by proper irrigation so that the product reaches the roots of the plants.

4.3.9.1 Application of Neem Soil Conditioner

Neem seed granules or powdered seeds are used to manufacture the soil conditioner, it can be applied during sowing of plants or can be sprinkled and raked into the soil. The process of sprinkling should be followed by proper irrigation so that the product reaches the roots of the plants.

4.3.10 Neem Biocontrol Agents

Biocontrol agents are nothing but pest control agents; neem is being used in agriculture to manufacture natural pest control agents for its safety, effectiveness and low cost. It is more effective than its synthetic counterparts. Properties of neem help to naturally control the pest growth rate. It is being commercially produced on large scale in organic farming and agriculture for best results.

4.3.10.1 Neem Biological Control

Over a decade farmers have realized the very fact that for effective biological control conservation for natural enemies of harmful pests is vital. There are different ways of habitat modification, they can be done in areas like Fields, Orchards, Vineyards, along or near the perimeters of fields, Hedges or uncultivated area. Actually, biological control is usage of useful living organism to control pest. This useful organism might be a predator or a parasite disease which attacks the harmful Insect or Pest. A biological control programme includes choosing a pesticide which is less harmful like Neem Pesticides or other Neem products. It also includes raising an insect and allowing it to attack another like a living insecticide.

4.3.10.2 Advantages

Biological control uses various kinds of useful predators and parasites hence they are able to reduce legal, environmental and chemical hazards. It is also economical. It also stops economic damage to agricultural crops. Biological Control is very pest-specific. Other useful insects remain unaffected by biological control. Environment and water remain unaffected.

4.3.10.3 Disadvantages

Biological Control is a time taking-process, includes a lot of data keeping, requires a lot of training and education, needs an intense learning of the pest and its enemies, biological control is costlier than pesticides. The results are not all that quick since it uses insects and parasites.

But usage of Neem helps in biological control as it complements the pests and predators and has no harmful effects on them.

4.4 Post Harvest Management/Miscellaneous

4.4.1 Storage Control

Neem has been traditionally been used for keeping stored grains free from pests and insects. According to estimates that almost 10% of total stored grains are lost every year due to pests and insects, the figures are much higher in developing countries, where post-harvest losses of grains are very high. Neem derivatives are being applied in all stores grains in farms and warehouses to protect the stored grains and prevent loss. Neem derivatives are more effective and economical as compared to their synthetic/chemical counterparts. The usage of neem lessens the dependence on costly pesticides and fumigants, thereby benefiting small and medium farmers.

4.4.1.1 Effectiveness of Neem Seed Kernel

Neem seed kernels mixed with rice/paddy helped to reduce damage to grain and reduce infestation over a period of three months. Use of neem to stored wheat also helped to treat insects and pests. Small and medium farmers obtain a high benefit-cost ratio by using neem to keep the stored grain pests at bay.

4.4.1.2 Effectiveness of Neem Oil and Neem Leaf

Neem oil and leaves have also been used to protect stored grains and legumes from major grain pests. Neem leaves are mixed with the grain in storage or the grain is stored in jute bags treated with neem oil or other neem extracts. These methods have effectively protected food and seed stores from insect pests for several months. Neem oil together with natural neem fumigant is said to have safely been used against five major stored grain pests that infest paddy and rice grains.

4.4.2 Livestock Feed

Neem tree has been used for animal feed as they contain amounts of nutrients, proteins, and minerals, required for animal health.

4.4.2.1 Neem Leaf

Contains sufficient amounts of proteins, amino acids, carotene, minerals except zinc. The fibre content is lower as compared to other leaves. They are also a source of providing sufficient amounts of nitrogen, potassium, and calcium. Animals like goat and camel are fed neem leaves as a part of their daily feed during winter. High quantity of digestible crude protein and total digestive nutrients makes it apt for being fed to cattle and buffaloes. Leaves as a feed have also led to weight gain in sheep and crossbred lambs. Chemical composition of neem leaf is moisture, protein, fat, fibre, carbohydrate, mineral, calcium, phosphorus, iron, thiamine, niacin, and vitamin C.

4.4.2.2 Seed Cake

Neem seed cakes are high in protein content so it used to reduce shortage of animal protein supplement in high producing animals. It contains all essential and nonessential amino acids, sulphur, etc. Though neem seed cake is bitter, but after water washing and other treatments helping in removing the bitter principles, it can be used as an animal feed for cattle on a regular basis. It has been an excellent source of cheap protein supplement for almost all kinds of livestock. Chemical composition of neem seed cake is moisture, carbohydrate, fat organic matter crude protein ether extract silica mineral phosphorous calcium potash. The waste remaining after extraction of the oil from neem seeds (neem seed cake) can be used as a biofertilizer, providing the macronutrients essential for plant growth (Ramachandran et al. 2007; Lokanadhan et al. 2012).

4.4.3 Soil Reclamation

The huge volume of sludge emanating from the tannery effluent treatment plants poses a serious environmental problem. Phytoremediation is an emerging technology in which the plants are employed to reclamate the contaminated soil strewn with heavy metals (metalloids) and toxic compounds. This work focuses the impact of application of tannery sludge on biochemical properties of 6 months old tree saplings of *Azadirachta indica* A. Juss. (Neem), *Melia azedarach* Linn. (Wild Neem), and *Leucaena leucocephala* (Lam) de Wit (Subabool) raised over the tannery sludge in an attempt to use these plants for phytoremediation. The plants raised over the garden soil served as the control. The porosity and water holding capacity of the tannery sludge were higher. The plant growth supporting elements such as Ca, total N_2, NO_3, and Mg were higher in the sludge. The plants raised over the sludge were found to be dark green with increased morphometric parameters. Electrophoretic profile revealed amplification of a few polypeptides (100, 105, 49, and 55 KDa). The levels of biomolecules and the CO_2 absorption increased in 6 months old plants. There was a significant uptake and transport of chromium in all the three tree species suggesting that these plants could be employed in phytoremediation of soils contaminated with heavy metals (Sakthivel and Vivekanandan 2009). A special type of red sandy dunal soil [Theri-soil] of Tamil Nadu is called Theri-soils. Theri-soils (Theri-soil) are located in Tuticorin, Tirunelveli, and Kanyakumari districts of Tamil Nadu. The merits of Theri lands are deep sand zone, good permeability, and quality groundwater. The demerits like the surface of the soil is not plane, higher level of soil erosion, sand dunes, from the top to the bottom only sand, low nutrients and minerals and Low water holding capacity makes them unsuitable for agriculture. Vermicompost and neem compost are the material used for the amendment of the Theri soils selected for the study to improve the fertility constraints of the soil. Measurements were made on the physicochemical and physical properties such as pH, EC, particle density, bulk density, porosity, water holding capacity, organic carbon content, and hydraulic conductivity. To convert this soil into a cultivable land, attempts were made to improve the soil moisture characteristics of the soil using soil amendment (Sundaram and Annadurai 2017).

4.5 Neem in Crop Management

Neem plays a vital role in crop and pest management. The use of neem in the form of neem fertilizers, insecticides, neem seed cake, neem pesticides, neem manure, neem soil conditioner, neem biocontrol agents, etc. treat problems of significant field crops, pests in field crops. Neem helps in controlling the growth of pests and insects in crops, thereby helps in overall increased productivity of agriculture. Neem is a well-known crop and pest manager, treating efficiently a large variety of crops and plants and also is an excellent insect/pest growth inhibitor. Every part or product of the neem tree (*Azadirachta indica*) (i.e., leaves, oils, stem, barks, oilcake, sawdust, etc.) have been shown to have a significant role to play in pest management. The different methods and techniques employed to obtain neem products can result in different concentrations of the active compounds, as well as different biological effectiveness (Roychoudhury 2016) (Table 4.5).

4.6 Impact on Environment

A major challenge of agriculture is to increase food production to meet the needs of the growing world population, without damaging the environment. In current agricultural practices, the control of pests is often accomplished by means of the excessive use of agrochemicals, which can result in environmental pollution and the development of resistant pests. In this context, biopesticides can offer a better alternative to synthetic pesticides, enabling safer control of pest populations. Attention is increasingly being paid to the use of natural compounds (such as essential oils) as a promising option to replace agrochemicals in agricultural pest control. These odoriferous substances are extracted from various aromatic plants, which are rich sources of biologically active secondary metabolites such as alkaloids, phenolics, and terpenoids (Esmaeili and Asgari 2015), using extraction methods employing aqueous or organic solvents, or steam distillation. Their mechanisms of action can vary, especially when the effect is due to a combination of compounds (de Oliveira et al. 2014; Esmaeili and Asgari 2015). Scientific research has shown that neem is safe for workers, with no handling risks, and can be used throughout the entire crop production cycle (Boeke et al. 2004. Neem has acquired commercial recognition due to its various beneficial properties, which have been extensively investigated over time. Compared to conventional chemicals, which are generally persistent in the environment and highly toxic, botanical pesticides are biodegradable and leave no harmful residues. Most botanical pesticides are non-phytotoxic and are also more selective toward the target pest. In terms of commercial applications, biopesticides can provide substantial economic advantages, since the infrastructure required is inexpensive, compared to conventional pesticides (Pant et al. 2016).

Table 4.5 Neem applications and commercial products available worldwide

Application	Product	Manufacturer
Fertilizer	Ozoneem cake ®	Ozone Biotech (India)
	Plan "B" organics-neem cake ®	Plan "B" Organics (USA)
	Fortuneem cake®	Fortune Biotech (USA)
	Bio neem oil foliar®	FUSA-Fertilizers of the USA
	Neem cake®	Unibell Corporation (Russia)
	Ozoneem coat®	Ozone Biotech (India)
	Parker neem coat®	Parker Neem (India)
	Neem urea guard®	Neemex (India)
	Fortuneem coat®	Fortune Biotech (USA)
Agrochemical	AZA-direct®	Gowan Company (USA)
	Neemex 4.5®	Certis (USA)
	Fortune aza 3% EC®	Fortune Biotech (USA)
	Azamax®	UPL Ltda. (Brazil)
	Neemazal technical®	E.I.D Parry Ltd. (India)
	Ecosense®	Agrologistics Systems Inc. (USA)
	Safer brand 3 in 1 garden spray®	Woodstream Corp. (Canada)
	Azatin XL®	OHP Inc. (USA)
	Azact CE®	EPP Ltda. (Brazil)

Source Campos et al. (2016)

The integrated use of botanical insecticides associated with biological control (synergism) in IPM is becoming increasingly widespread in the farming and research communities. The advantage of this approach is that it offers the potential to control agricultural pests, without serious impacts on the environment, nontarget organisms, and animal and human health. Other factors that have stimulated the use of neem-based products for pest control in agriculture are ecological and toxicological aspects (low toxicity to non-target organisms), as well as economic aspects (small amounts of the product can provide effective pest control (Ogbuewu et al. 2011).

4.7 Conclusion and Future Thrust

Scientists across the globe are anticipating that revival of large-scale cultivation of neem tree cultivation would create a new era in the pest control and has a very high potential in mitigating the current ecological problems which could eventually lead us to a sustainable development. Recently, more emphasis has been given towards the neem as an organic alternative to industrial pesticides. Neem-based pesticides are the major alternatives to the conventional synthetic chemical pesticides due to various advantages over conventional pesticides. It is essential to prevent the development of pest resistance to the botanical pesticides or by using different botanicals periodically so that pest would not able to recognize the compounds and develop resistance mechanism against the botanical pesticides. Despite many promising properties of neem-based biopesticides, there are limitations that hinder effective large-scale use of neem including short shelf life, photosensitivity, and volatilization, make it difficult to use them on a large scale. These impediments must be overcome and many uncertainties should be clarified so that the full potential of neem can be exploited. However, lack of industrial interest is one of the main problems facing the commercial development of neem, largely due to the difficulty of patenting natural products, as well as a shortage of scientific evidence to support claims regarding the benefits of these substances. As a result, the products are not widely publicized in the farming community. Thus, there is a huge possibility of exploring the more potential of neem and its products in protecting crops in a cost-effective and eco-friendly way.

References

Allameh R, Razzaq B, Razzaghi M, Shams M, Rezaee MB, Jaimand K (2002) Effect of neem leaf extract on production of aflatoxin and activities of fatty acid synthetase, isocitrate dehydrogenase and glutathione S-transferase in *Aspergillus parasiticus*. Mycopathologia 154:79–84

Anis Joseph R, Premila KS, Nisha VG, Rajebdran S, Mohan SS (2010) Safety of neem products to tetragnathid spiders in rice ecosystem. J Biopesticides 3(1):88–89

Anonymous (1997) Neem foundation 1997. http://www.neemfoundation.org

Anonymous (2009) Cold pressed neem oil; exemption from the requirement of a tolerance. Federal Register, Environmental Protection Agency, USA

Biswas K, Chattopadhya I, Banerjee RK, Bandopadhyay U (2002) Biological activities and medicinal properties of neem (Azadiracta indica A. Juss). Curr Sci 82(11):10

Boeke SJ, Boersma MG, Alink GM, van Loon JJ, van Huis A, Dicke M, Rietjens IM (2004) Safety evaluation of neem (*Azadirachta indica*) derived pesticides. J Ethnopharmacol 94:25–41

Brahmachari G (2004) Neem—an omnipotent plant: a retrospection. ChemBioChem 5:408–421

Campos EVR, de Oliveira JL, Pascoli M, de Lima R, Fraceto LF (2016) Neem oil and crop protection: from now to the future. Front Plant Sci 7:1494

Chandler D, Bailey AS, Tatchell GM, Davidson G, Greaves J, Grant WP (2011) The development, regulation and use of biopesticides for integrated pest management. Philos Trans R Soc B Bio Sci 366(1573):1987–1999

Cooper J, Dobson H (2007) The benefits of pesticides to mankind and the environment. Crop Prot 26:1337–1348

Cox C (2002) Prethrins/pyrethrum insecticide fact sheet. J Pesticides Reform 22:14–20

Damalas CA, Eleftherohorinos IG (2011) Pesticide exposure, safety issues, and risk assessment indicators. Int J Environ Res Public Health 8(5):1402–1419

Doutt RL (1964) The historical development of biological control. In: Debach P (ed) Biological control of insect pests and weeds. Reinhold Publishing Corporation, New York, NY

De Oliveira JL, Campos EVR, Bakshi M, Abhilash PC, Fraceto LF (2014) Application of nanotechnology for the encapsulation of botanical insecticides for sustainable agriculture: prospects and promises. Biotechnol Adv 32:1550–1561

Esmaeili A, Asgari A (2015) In vitro release and biological activities of *Carum copticum* essential oil (CEO) loaded chitosan nanoparticles. Int J Biol Macrom 81:283–290

Ghorbanian M, Razzaghi-Abyaneh M, Abdolamir A, Masoomeh SG, Qorbani M (2007) Study on the effect of neem (*Azadirachta indica* A. Juss) leaf extract on the growth of *Aspergillus parasiticus* and production of aflatoxin by it at different incubation times. Mycoses 51:35–39

Gonzalez-Coloma A, Reina M, Diaz CE, Fraga BM, Santana-Meridas O (2013) Natural product-based biopesticides for insect control. In: Reedijk J (ed) Reference module in chemistry, molecular sciences and chemical engineering. Elsevier, Amsterdam, pp 1–53

Harjai K, Bala A, Gupta RK, Sharma R (2013) Leaf extract of *Azadirachta indica* (neem): a potential antibiofilm agent for *Pseudomonas aeruginosa*. Pathogens Dis 69:62–65

Hossain MA, Nagooru MR (2011) Biochemical profiling and total flavanoids contents of leaves crude extract of endemic medicinal plant Cordyline terminalis L. Kunth. Pharmacogn J 3:25–30

Islam MT, Omar D, Latif MA, Morshed MM (2011) The integrated use of entomopathogenic fungus, *Beauveria bassiana* with botanical insecticide, neem against *Bemisia tabaci* on eggplant. Afr J Microbiol Res 5:3409–3413

Islam MT, Omar DB (2012) Combined effect of *Beauveria bassiana* with neem on virulence of insect in case of two application approaches. J Anim Plant Sci 22:77–82

Koul O, Singh G, Singh R, Singh J, Daniewski WM, Berlozecki S (2004) Bioefficacy and mode of action of some limonoids of salannin group from Azadirachta indica A. Juss and their role in a multicomponent system against lepidopteran larvae. J Biosci 29:409–416

Lokanadhan S, Muthukrishnan P, Jeyaraman S (2012) Neem products and their agricultural applications. J Biopesti 5:72–76

Mohanty S, Patra AK, Chhonkar PK (2008) Neem (Azadirachta indica) seed kernel powder retards urease and nitrification activities in different soils at contrasting moisture and temperature regimes. Bioresour Technol 99(4):894–899

Mamoon-ur-Rashid M, Khattak MK, Abdullah K, Hussain S (2011) Toxic and residual activities of selected insecticides and neem oil against cotton mealybug, phenacoccus solenopsis tinsley (sternorrhyncha: pseudococcidae) under laboratory and field conditions. Pak Entomol 33:151–155

Marz U (1989) The economics of neem production and its use in pest control. Wissenschaftsverlag Vauk Kiel

Moslem MA, El-Kholie EM (2009) Effect of neem (*Azardirachta indica* A. Juss) seeds and leaves extract on some plant pathogenic fungi. Pak J Biol Sci 12:1045–1048

Mossini SAG, Kemmelmeier C (2005) A árvore Nim (*Azadirachta indica* A. Juss): múltiplos usos. Acta Farm Bonaer 24:139–148

Mueller DS, Dorrance AE, Derksen RC, Ozkan E, Kurle JE, Grau CR, Gaska JM, Hartman GL, Bradley CA, Pedersen WL (2002) Efficacy of fungicides on *Sclerotinia sclerotiorum* and their potential for control of *Sclerotinia* stem rot on soya bean. Plant Dis 86:26–31

Musalia L, Anandan S, Sastry VR, Agrawal D (2000) Urea-treated neem (*Azadirachta indica* A. juss) seed kernel cake as a protein supplement for lambs. Small Rumin Res 35:107–116

Ni K, Pacholski A, Kage H (2014) Ammonia volatilization after application of urea to winter wheat over 3 years affected by novel urease and nitrification inhibitors. Agric Ecosyst Environ 197:184–194

Nicoletti M, Petitto V, Gallo FR, Multari G, Federici E, Palazzino G (2012) The modern analytical determination of botanicals and similar novel natural products by the HPTLC fingerprint approach. Stud Nat Prod Chem 37:217–258

Niu JZ, Hull HS, Zhang YX, Lin JZ, Dou W, Wang JJ (2014) Biological control of arthropod pests in citrus orchards in China. Biol Control 68:15–22

Oerke EC, Dehne HW (2004) Safeguarding production-losses in major crops and the role of crop protection. Crop Prot 23:275–285

Ogbuewu IP, Odoemenam VU, Obikaonu HO, Opara MN, Emenalom OO, Uchegbu MC, Okoli IC, Esonu BO, Iloeje MU (2011) The growing importance of neem (*Azadirachta indica* A. Juss) in agriculture, industry, medicine and environment: a review. Res J Med Plant 5:230–245

Pant M, Dubey S, Patanjali PK (2016) Recent advancements in bio-botanical pesticide formulation technology development. In: Veer V, Gopalakrishnan R (eds) Herbal insecticides, repellents and biomedicines: effectiveness and commercialization. Springer, New Delhi, pp 117–126

Ploetz RC (2000) Panama disease: a classic and destructive disease of panama. Plant Health Prog. https://doi.org/10.1094/php-2000-1204-01-hm

Raguraman S, Kannan M (2014) Non-target effects of botanicals on beneficial arthropods with special reference to Azadirachta indica. In D. Singh (ed) Advances in plant biopesticides. Springer, New Delhi, pp 173–205

Ramachandran S, Singh SK, Larroche C, Soccol CR, Pandey A (2007) Oil cakes and their biotechnological applications-a review. Bioresour Technol 98(10):2000–2009

Ramos AR, Falcao LL, Barbosa GS, Marcellino LS, Gander ES (2007) Neem (*Azadirachta indica* A. Juss) components; candidates for the control of *Crinipellis perniciosa* and *Phytophthora* spp. Microbiol Res 162:238–243

Rajendran SM (2010) Bioefficacy of neem insecticidal soap (NIS) on the disease incidence of bhendi, Abelmoschus esculentus (L.) Moench under field conditions. J Biopesticides 3(1):246

Roychoudhury R (2016) Neem products. In: Omkar (ed) Ecofriendly pest management for food security. Elsevier, Amsterdam, pp 545–562

Sakthivel V, Vivekanandan M (2009) Reclamation of tannery polluted soil through phytoremediation. Physiol Mol Biol Plants Int J Funct Plant Biol 15(2):175–180

Sarup P, Srivastava VS (1971) Observations on the damage of neem (Azadirachta indica A. Juss) seed kernel in storage by various pest and efficacy of the damaged kernel as an antifeedant against the desert locust, Schistocerca gregaria F. Indian J Entomol 33:228–230

Schmutterer H (1990) Properties and potential of natural pesticides from the neem tree, Azadirachta indica. Annu Rev Entomol 35:271–297

Sola P, Mvumi M, Ogendo JO, Mponda O, Kamanula JF, Nyirenda SP, Belmain SR, Stevenson PC (2014) Botanical pesticide production, trade and regulatory mechanisms in sub-Saharan Africa: making a case for plant based pesticidal products. Food Secur 6:369–384

Subapriya R, Nagini S (2005) Medicinal properties of neem leaves: a review. Current Med Chem—Anticancer Agents 5:146–149

Sujarwo W, Keim AP, Caneva G, Toniolo C, Nicoletti M (2016) Ethnobotanical uses of neem (Azadirachta indica A. Juss.; Meliaceae) leaves in Bali (Indonesia) and the Indian subcontinent in relation with historical background and phytochemical properties. J Ethnopharmacol 189:186–193

Sundaram S, Annadurai B (2017) Soil reclamation using vermicompost and neem compost amended theri soil in tuticorin district. Int J Develop Res 7 (10):15746–15748

Shanmugam V, Kumar S, Singh MK, Verma R, Sharma V, Ajit NS (2007) First report of alstroemeria wilt caused by *Fusarium oxysporum* in India. Plant Pathol 56:727

Sturrock CJ, Woodhall J, Brown M, Walker C, Mooney SJ, Ray RV (2015) Effects of damping-off caused by *Rhizoctonia solani* anastomosis group 2-1 on roots of wheat and oil seed rape quantified using X-ray computed tomography and real-time PCR. Front Plant Sci 6:461

Tunca H, Kilincer N, Ozkan C (2012) Side-effects of some botanical insecticides and extracts on the parasitoid, Venturia canescens (Grav.) (Hymenoptera: Ichneumonidae). Türk Entomol Derg 36:205–214

Vethanayagam SM, Rajendran SM (2010) Bioefficacy of neem insecticidal soap (NIS) on the disease incidence of bhendi, Abelmoschus esculentus (L.) Moench under field conditions. J Biopesticides 3(1):246–249

Venkateshwarlu B, Pirat M, Kishore N, Rasul A (2008) Mycorrihzal inoculation in neem (Azadiraxcta Indica) enhances azadiractin content in seed kernels. World J Microbio Biotechnol 24:1243–1247

Wilps H, Kirkilionis E, Muschenich K (1992) The effects of neem oil and azadirachtin on mortality, and energy metabolism of Schistocera gregaria forskal—a comparison between laboratory and field locusts. Pharmacol 102:67–71

Young CS, Clarkson JP, Smith JA, Watling M, Philips K, Whips JM (2004) Environmental conditions influencing *Sclerotinia sclerotiorum* infection and disease development in lettuce. Plant Pathol 53:387–397

Phylogeny of Neem and Related Species in the Meliaceae Family

Nagesh A. Kuravadi and Malali Gowda

Abstract

Neem belongs to Meleaceae family also called Mahogany family. It includes 50 genera and 1400 species. In India, it is represented by 20 genera and 70 species. However, the chemical composition of these species has high variability. This study was aimed at understanding the genetic distance of neem tree in relation to other family members (*Azadirachta indica, Amoora lawii, Dysoxylum malabaricum, Khaaya, Melia Azadirachta, Melia dubia, Swietenia macrophylla, Sweitenia mahogany, Soymida fabrifuga, Toona ciliate, Walsuria trifoliate*). Additionally, species were also analysed for Azadirachtin concentration in the leaf with the aim to find clues about the evolution of Azadirachtin pathway and the family members who share it. Neem tree synthesises a large number of unique bioactive compounds. Chemical synthesis of many of these compounds has been tried in the laboratory; complex synthesis process and molecular complexity involved makes the process not a viable option for commercial production (Veitch et al. 2008). However, to increase the natural production or engineering new production methods require a deeper understanding of the synthesis process and also the genes/enzymes involved in their biosynthesis. Sequencing of neem genome is aimed at elucidating all the genes that are present in neem and understanding pathways of synthesis for many of the compounds that are produced by neem tree. Studying neem genome will also be useful in understanding the mechanism of resistance to abiotic stress.

5.1 Introduction

Neem tree is native to Indian sub-continent, This multipurpose tree is known to harbour wide-spectrum of biological activities. Over 200 secondary metabolites belong to a class isoprenoids including azadirachtin, nimbin and salanin (Breuer et al. 2003; Chattopadhyay et al. 2004) are found in neem. Azadirachtin is the most prominent bio-pesticide naturally found in neem tree (Kuravadi et al. 2015). The compound is naturally synthesised in secretory cells of neem tree which is the most abundant chemical in seed (Dayanandan et al. 2000). In addition, neem is known to contains antifungal (Aparecida Galerani Mossini et al. 2004), antibacterial (Chopra et al. 1952) and antiviral activities against Vaccinia, Chikungunya and measles viruses. Neem

N. A. Kuravadi
Centre for Cellular and Molecular Platforms,
National Centre for Biological Sciences, Bengaluru, Karnataka, India
e-mail: alwaysnagesh@gmail.com

M. Gowda (✉)
Center for Functional Genomics and Bio-Informatics, the University of TransDisciplinary and Health Sciences, Bengaluru, India
e-mail: malalig@tdu.edu.in

bioactive compounds are varied significantly depending on the genetic and environmental factors (Sidhu et al. 2003).

Neem belongs to Meliaceae family with many species. Miliaceae is a diverse family of trees with high amount of variability in both phenotypic and biochemical characteristics. Among the Meliaceae members only a few species can produce complex limonide compound like Azadirachtin. We sequenced the rbcL gene from various members of Meliaceae to understand the diversity and the biochemical pathways they share. Following is the methodology used for the analysis.

5.2 Materials Used

Meleaceae trees used for the study:

Sl. no.	Plant name
1.	*Azadirachta indica A. Juss* (Neem)
2.	*Amoora lawii*
3.	*Dysoxylum malabaricum*
4.	*Khaaya*
5.	*Melia Azadirachta*
6.	*Melia dubia*
7.	*Swietenia macrophylla*
8.	*Sweitenia mahogany*
9.	*Soymida fabrifuga*
10.	*Toona ciliata*
11.	*Walsuria trifoliata*

5.3 PCR Amplification of rbcL Gene

The PCR amplification was performed using rbcL specific primers to amplify the gene. Following primers were used for amplification:

Primer name	Primer sequences (5′–3′)
rbcL F	TCATGGTATGCACTTTCGTG
rbcL R	CTCCCATTTGCTAGCCTCACG

PCR master mix was prepared as follows:

PCR component	Per reaction (μl)
Dream taq buffer (10×)	5
dNTPs (5 mM)	2
Primers rbcL F (10 μM)	1.5
Primers rbcL R (10 μM)	1.5
Template DNA (10 ng/μl)	5
Taq polymerase (5U/μl)	0.2
Nuclease free water	34.8

5.3.1 PCR Amplification Cycle for rbcL Amplification

Initial denaturation step was set at 95 °C for 3 min, followed by 35 cycles of denaturation at 95 °C for 30 s, annealing at 55 °C for 45 s, extension at 72 °C for 90 s. Final extension was carried out at 72 °C for 5 min and hold was set at 4 °C.

5.4 Species Diversity Analysis

The rbcL gene sequence from 11 species of Meliaceae plants were sequenced along with Citrus. All the sequences were used in multiple sequence alignment using ClustalW followed by Dendrogram generation using MEGA5 using neighbor-joining method with bootstrapping of 1000 (Tamura et al. 2011). Leaf images of each tree are aligned against the species name in dendrogram to show the diversity of Meliaceae trees. Figure 5.1 shows the dendrogram of Meliaceae trees and the box highlighted shows the trees that produce Azadirachtin. Figure 5.1 shows the leaf images of Meliaceae plants that were compared based on DNA barcoding gene rbcL.

5.5 Conclusions

The study showed that the *C. maximus* from Rutaceae acted as an outgroup in the phylogeney and closer clustering of Meliaceae members. It also

5 Phylogeny of Neem and Related Species in the Meliaceae Family

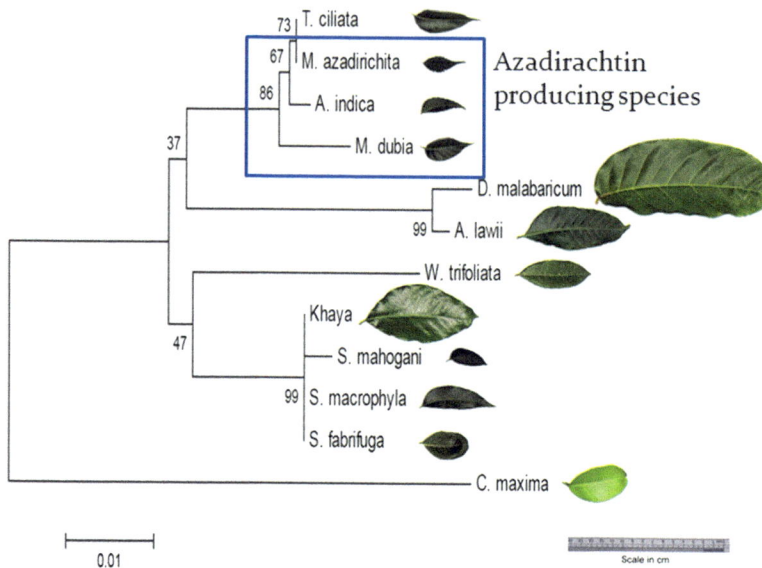

Fig. 5.1 Dendrogram along with leaf image of Meliaceae plants based on rbcL gene sequence

showed that neem falling into a small cluster of four species of which three were having Azadirachtin in them which suggested that the Azadirachtin pathway has been developed very recently in evolutionary history. It also provided an clue that the Azadirachtin production could be through evolution of a gene involved in limonide bio-synthetic pathway shared by all Meleaceae trees.

Acknowledgements We acknowledge Genomics facility (BT/PR3481/INF/22/140/2011) at Centre for Cellular and Molecular Platforms, Bangalore for sequencing of Neem genomes. We acknowledge Pradeep H, Aarati Karaba, Manojkumar S and Annapurna for their help in NGS library preparation and sequencing. We thank Ashmita G and Divya S for their help in manual curation of SSR markers. We are grateful to Rajanna, National Botanical Garden, University of Agricultural Sciences, GKVK campus, Bangalore for his help during neem sample collection.

References

Aparecida Galerani Mossini S, De Oliveira KP, Kemmelmeier C (2004) Inhibition of patulin production by Penicillium expansum cultured with neem (Azadirachta indica) leaf extracts. J Basic Microbiol 44 (2):106–113

Breuer M, Hoste B, De Loof A, De Naqvi SNH (2003) Effect of Melia azedarach extract on the activity of NADPH-cytochrome c reductase and cholinesterase in insects. Pestic Biochem Physiol 76:99–103

Chattopadhyay I, Nandi B, Chatterjee R, Biswas K, Bandyopadhyay U, Banerjee RK (2004) Mechanism of antiulcer effect of neem (Azadirachta indica) leaf extract: effect on $H + -K + -ATPase$, oxidative damage and apoptosis. Inflammopharmacology 12:153–176

Chopra IC, Gupta KC, Nazir BN (1952) Preliminary study of anti-bacterial substances from Melia azidirachta. Indian J Med Res 40:511

Dayanandan P, Ponsamuel J, Gupta PD, Yamamoto H (2000) Ultrastructure of terpenoid secretory cells of neem (Azadirachta indica A. Juss.). In: Electron microscopy in medicine and biology, p 179–195

Kuravadi NA, Yenagi V, Rangiah K, Mahesh HB, Rajamani A, Shirke MD, Russiachand H, Loganathan RM, Lingu CS, Siddappa S (2015) Comprehensive analyses of genomes, transcriptomes and metabolites of neem tree. PeerJ 3:e1066

Sidhu O, Kumar V, Behl H (2003) Variability in neem (Azadirachta indica) with respect to azadirachtin content. J Agric Food Chem 51:910–915

Tamura K, Peterson D, Peterson N, Stecher G, Nei M, Kumar S (2011) MEGA5: molecular evolutionary genetics analysis using maximum likelihood, evolutionary distance, and maximum parsimony methods. Mol Biol Evol 28:2731–2739

Veitch GE, Boyer A, Ley SV (2008) The azadirachtin story. Angew Chem Int Ed 47:9402–9429

Strategies and Tools for Next Generation Sequencing

Nagesh A. Kuravadi and Malali Gowda

Abstract

Next Generation Sequencing (NGS) technologies have accelerated the process of whole genome sequencing. It provided a faster and easier way to sequence whole genomes at a fraction of cost and time. With the availability of multiple technologies; choices still needs to be made when planning a genome sequencing project in order to maximize assembly quality versus cost. For neem genome sequencing we have used Illumina and Roche sequencing technologies. To sequence and assemble neem genome, we isolated high quality genome DNA from neem leaves and library was prepared for paired end DNA sequencing using Illumina HiSeq 1000. Library was also prepared for Roche 454 sequencer to generate long reads from genome. Transcriptome sequence from RNA of various tissues and seedlings was also generated to assesses the expressed genes and support the gene prediction and annotation process.

N. A. Kuravadi
Centre for Cellular and Molecular Platforms,
National Centre for Biological Sciences, Bengaluru,
Karnataka, India
e-mail: alwaysnagesh@gmail.com

M. Gowda (✉)
Center for Functional Genomics and
Bio-informatics, The University of
TransDisciplinary and Health Sciences,
Bengaluru 560064, India
e-mail: malalig@tdu.edu.in

6.1 Introduction

Traditionally genomes sequencing has been a challenging task because of slow and labour intensive methods like Sanger's di-deoxy sequencing system and lack of bioinformatic applications. However, Next Generation Sequencing (NGS) technologies have accelerated the process of whole genome sequencing and provided an opportunity to sequence whole genomes at a fraction of cost and time. Despite the dramatic drop in sequencing cost and availability of various technologies, we still need to be careful in planning a genome project in order to maximize assembly quality versus cost (Li and Harkess 2018).

Not all plants are easily sequenced using NGS technology. Various factors that affect the feasibility of the project like genome size, repeat structure and zygosity of the species being sequenced (Li and Harkess 2018). These factors affect the choice of NGS methods/technologies that have their own pros and cons.

Being a non-model tree we had limited background information on neem genome. Neem is a diploid tree. The chromosome number of neem is under debate; $2n = 28$ (Deshmukh 1959) and $2n = 2x = 24$ (Singh and Chaturvedi 2013). Only one report is available for neem genome size estimation, which showed 384 Mb as the genome size of neem (Ohri et al. 2004).

Based on the available information, we had chosen two different technologies to complement

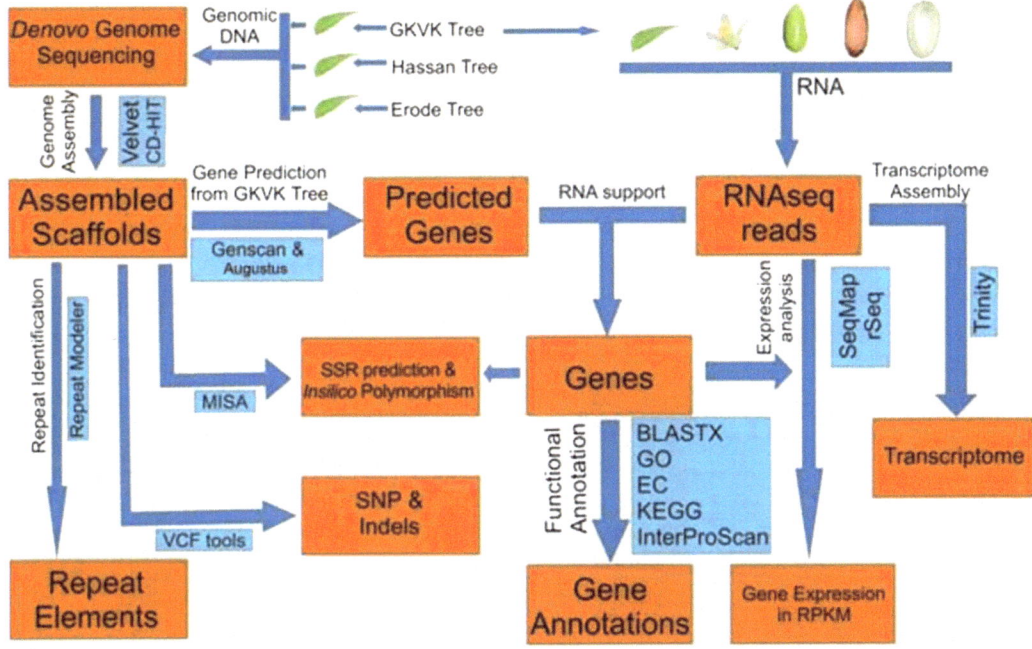

Fig. 6.1 Work flow of neem genome and transcriptome analysis

each other to generate whole genome assembly of neem (Kuravadi et al. 2015). We have also complemented the genome sequencing effort with sequencing of transcriptome of multiple tissues to support gene prediction and expression analysis. Figure 6.1 describes the strategies and workflow used for sequencing neem genome.

6.2 Methodology

See Fig. 6.1.

6.2.1 Genomic DNA Isolation from Neem Leaves

DNA isolation from Neem was optimized by using a combination of CTAB method and Sigma Genelute plant DNA isolation kit. For DNA isolation 1–2 g of fresh leaves were ground along with 6 ml of 6% CTAB buffer (6% CTAB, 1.4 M NaCl, 20 mM EDTA, 10 mM Tris Base pH8). The mixture was divided into multiple 1.5 ml tubes such that ∼750 μl was present in each tube. 7.5 μl of Beta-Mercaptoethanol (1% of the volume in the tubes) was added and the tube was vortexed thoroughly for 15 s. The tube was then incubated at 65 °C for an hour and mixed by inverting every 15 min. The tubes were then centrifuged at 14,000 g for 15 min. The supernatant was taken in a new eppendorf tube and 750 μl of Chloroform:iso amyl alcohol mixture (24:1) was added to it. The tubes were then mixed by inverting for a minute and centrifuged at 14,000 g for 20 min. The supernatant aqueous phase was again taken in a new eppendorf tube and 750 μl of Chloroform:iso amyl alcohol mixture (24:1) was added to it. The tubes were then mixed by inverting for a minute and centrifuged at 14,000 g for 20 min. The supernatant was taken in a new eppendorf tube and purified using the Sigma Genelute plant DNA isolation kit (G2N70, Sigma). For every 200 μl of supernatant 350 μl Lysis buffer A, 50 μl Lysis buffer B was added, mixed and incubated at 65 °C for 10 min. 130 μl of precipitation buffer was added and incubated on ice for 5 min. The tube was then centrifuged at 14,000 g for 1 min. The supernatant was removed and passed through the filtration column. To the filtrate 700 µl of binding solution was added. Nucleic acid binding column was prepared with 500 µl of column preparation solution and centrifuging at 12,000 g for 1 min, discarding the flow through. To the prepared column 500 µl of the filtrate with the binding solution was then added to the binding column and centrifuged at 12,000 g for 1 min and the flow through discarded. 500 µl of wash solution was added to the column and centrifuged at 12,000 g for 1 min and the same step repeated again. The column was then transferred to a new tube. The DNA was then eluted by adding 100 µl of elution buffer (prewarmed at 65 °C) centrifuging at 12,000 g for 1 min. The DNA was run on a 1% agarose gel. The A260/280 ratio and the quantity was determined using the nanodrop. We used Illumina paired end sequencing to generated large amount of short read data that is highly accurate and also provide paring information. We prepared DNA library for three neem Genotypes (1, 2 and 3).

6.2.2 Paired End Illumina Sequencing

Libraries were prepared using TrueSeq DNA sample preparation kit (FC-121-2001) from Illumina (www.illumina.com). In brief, the protocol involved fragmentation of DNA using covaris focused-ultrasonicator to fragment the DNA to the required size. The fragmented DNA was end repaired and A tailed followed by Illumina adapter ligation. The libraries were checked for quality on bioanalyzer and quantified using qubit fluorometer. The QC passed libraries were sequencing on Illumina HiSeq 1000 using paired-end (PE) (2 × 100 nts) sequencing chemistry (Fig. 6.2) at Next Generation Genomics Facility at Centre for Cellular and Molecular Platforms (C-CAMP). We generated 13.86 Giga bases (Gb), 8.70 Gb and 17.30 Gb of high quality Illumina data for Genotypes 1, 2 and 3, respectively.

6.2.2.1 Roche 454 Library Preparation and Sequencing

We also sequenced whole genome of neem using Roche 454 sequencing chemistry. Roche produces long single end reads that is also highly accurate. However, the data output from the machine is very low when compared to Illumina sequencer. To obtain long read data we prepared 454 library from neem Genotype 1 (GKVK, Bangalore) using the rapid library preparation kit from Roche (Cat. No. 05608228001y; version 4.0.12). In brief, the protocol involved in fragmentation of DNA by nebulisation of sample in a cup by applying 30 psi pressure of nitrogen. Fragmented DNA is end repaired and adapter legated. The final library is quantitated using qubit and quality was assessed using Agilent bioanalyzer. QC passed library was sequenced using GS FLX+ chemistry as per Roche manual instructions (http://454.com) (Fig. 6.3).

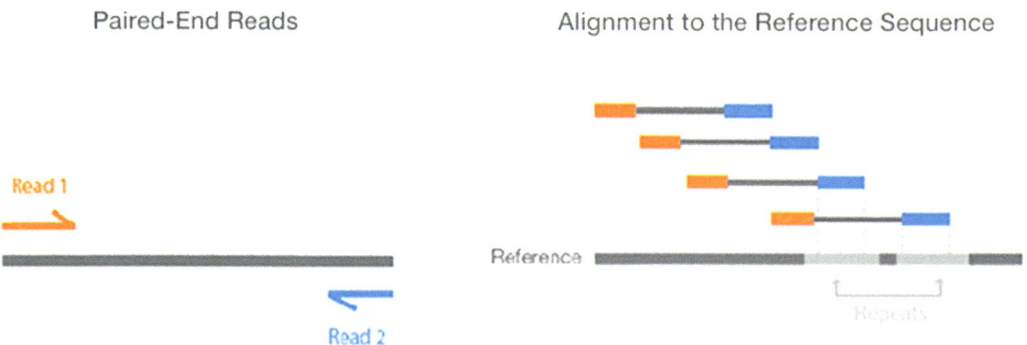

Fig. 6.2 Sequence reading in paired-end sequencing format (https://www.illumina.com/science/technology/next-generation-sequencing/paired-endvs-single-read-sequencing.html)

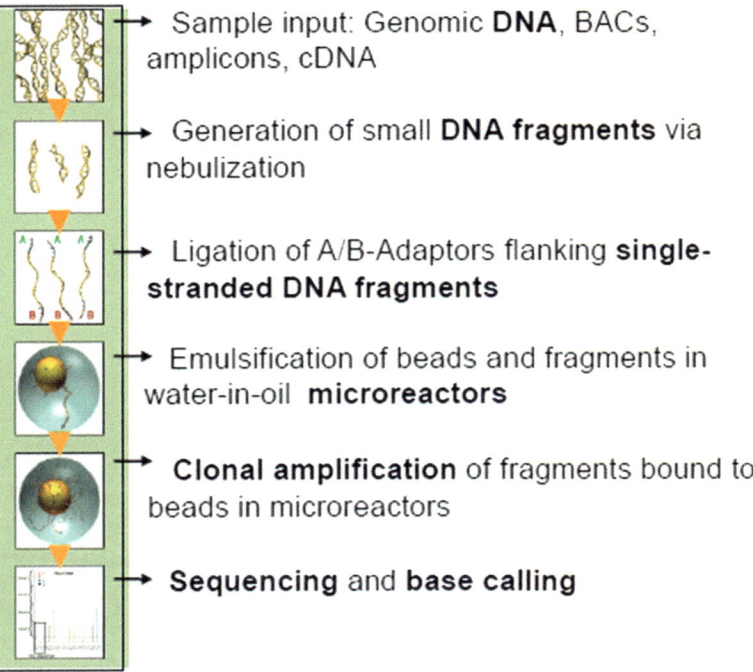

Fig. 6.3 Workflow for pyrosequencing (http://symposcium.com/2014/01/qa-what-is-the-workflow-of-454-pyrosequencing)

We generated 1.13 Gb of 454 long reads (500–1000 nts) for Genotype 1. The average Roche read length was 410 nts and longest read length of 1596 nts.

6.2.3 Transcriptome Sequencing

Apart from whole genome data, we also sequenced transcriptome from five different tissues (Leaf, Flower, Developing endosperm, Fruit coat with pulp and Mature fruit) of neem tree Genotype 1. Also, transcriptome sequencing of different seedling under normal and drought conditions were performed.

Total RNA isolation was carried out with 100 mg of the plant tissue using the Sigma Spectrum Plant Total RNA Kit (STRN50, Sigma, Seelze, Germany). The Tissue was ground using liquid nitrogen in a mortar and pestle. The sample was then taken in an eppendorf tube and 500 µl of lysis solution with 5 µl of Beta-mercapto ethanol was added to the sample and vortexed immediately for at least 30 s. The samples were then incubated at 56 °C for 3–5 min. The sample was then centrifuged at 14,000 g for 3 min to pellet cellular debris. The lysate was then added to the filtration column and centrifuged at 14,000 g for 1 min. To the clarified filtrate 500 µl of binding solution was added and mixed immediately by pipetting up and down 5 times. 700 µl of the sample was passed through a binding column and the flow through was discarded. DNase treatment was done to the RNA that bound on the column using the Sigma On-column DNase digestion kit (DNASE 70, Sigma). For this 300 µl of Wash solution was added to the column and centrifuged at 14,000 g for 1 min. The flow through was decanted. To the column a mixture of and 10 µl of DNase 1 and 70 µl of buffer was added to the centre of the binding column and incubated at room temperature for 1 min. 500 µl of Wash solution 1 was then pipetted and centrifuged at 14,000 g for 1 min. After the flow through was discarded 500 µl of Wash solution 2 was added to the column and centrifuged at 14,000 g for 1 min. The above step was repeated again. After the flow through was discarded the column was centrifuged again at 14,000 rpm to get rid of any residual wash solution. For the RNA elution the column was transferred to a new eppendorf tube

Table 6.1 Raw reads obtained for tissues by RNAseq

Sample	No. raw reads
Mature Leaf (MAL)	15,766,151
Bud and Flower (BAF)	65,374,167
Green Fruit Coat and Pulp (GFP)	159,821,489
Embryo and Endosperm (EAE)	88,054,962
Dry Seed (DYS)	69,999,926
Callus (CAL)	6,183,186
Normal Root (NRT)	12,396,503
Normal Shoot (NST)	16,764,704
Drought Root (DRT)	20,728,964
Drought Shoot (DST)	63,114,989
Albino Root (ART)	36,134,685
Albino Shoot (AST)	65,729,213

and 50 μl of DEPC treated water was added to the column. The column was allowed to sit for a minute and then centrifuged at 14,000 g for 1 min. The RNA quality was determined using the Agilent Bioanalyzer. The RIN (RNA integrity value) values of all the samples were greater than 6. The A260/280 ratio and the quantity were determined using the nanodrop.

mRNA libraries were prepared using 1 μg of total RNA according to Illumina's TrueSeq RNA sample preparation kit (RS-122-2001). Library QC was done and sequencing was done on Illumina HiSeq 1000 using paired-end (PE) (2 × 100 nts) sequencing chemistry. The number of transcript reads obtained for each of tissue samples is summarized in Table 6.1.

6.3 Conclusion

This chapter explains the experimental design for sequencing genome of multiple genotypes of neem tree and sequence transcriptome of various tissues. Also explained in detail the optimized methods needed to isolate good quality DNA and RNA from neem tree. The library preparation and NGS sequencing of the samples generated large amount of data needed to assemble the genome and predict and validate the genes expressed in neem tree.

Acknowledgements We acknowledge Genomics facility (BT/PR3481/INF/22/140/2011) at Centre for Cellular and Molecular Platforms, Bangalore for sequencing of Neem genomes. We acknowledge Pradeep H, Aarati Karaba, Manojkumar S and Annapurna for their help in NGS library preparation and sequencing. We thank Ashmita G and Divya S for their help in manual curation of SSR markers. We are grateful to Rajanna, National Botanical Garden, University of Agricultural Sciences, GKVK campus, Bangalore for his help during neem sample collection.

References

Deshmukh NY (1959) Chromosome number in neem. Indian Oil Seeds J 3

Kuravadi NA, Yenagi V, Rangiah K, Mahesh HB, Rajamani A, Shirke MD, Russiachand H, Loganathan RM, Lingu CS, Siddappa S (2015) Comprehensive analyses of genomes, transcriptomes and metabolites of neem tree. PeerJ 3:e1066

Li FW, Harkess A (2018) A guide to sequence your favorite plant genomes. Appl Plant Sci 6:E1030

Ohri D, Bhargava A, Chatterjee A (2004) Nuclear DNA amounts in 112 species of tropical hardwoods—new estimates. Plant Biol 6:555–561

Singh M, Chaturvedi R (2013) Somatic embryogenesis in neem (*Azadirachta indica* A. Juss.): current status and biotechnological perspectives

Neem Genome Assembly

Nagesh A. Kuravadi and Malali Gowda

Abstract

Genome assembly refers to the process of taking a large number of short DNA sequences and putting them back together to create a representation of the original chromosomes from which the DNA originated. For assembling neem genome, we filtered the NGS raw reads from Illumina to get high-quality reads followed by assembly using Velvet assembler. Similar process was also followed to assemble Roche/454 reads using MIRA. The individual assembly from velvet and MIRA was merged using cd-hit to get the final assembly of neem genome. Transcriptome of neem was assembled from short-read data from different tissues using Trinity assembler. The reads for chloroplast and mitochondria were extracted separately by mapping genome reads to chloroplast and mitochondrial genomes of known plants. The reads were assembled separately using Velvet.

N. A. Kuravadi
Centre for Cellular and Molecular Platforms,
National Centre for Biological Sciences,
Bengaluru, Karnataka, India
e-mail: alwaysnagesh@gmail.com

M. Gowda (✉)
Center for Functional Genomics and Bio-Informatics, The University of TransDisciplinary and Health Sciences,
Bengaluru, India
e-mail: malalig@tdu.edu.in

7.1 Illumina Sequence Data Assembly

For a good assembly of genome, it is important the sequencing data input is of very high quality. Hence, the raw data obtained from Illumina sequencing were analysed for their quality and low-quality bases were filtered before assembling. Illumina PE reads were pre-processed using FASTX-Toolkit (v 0.0.13). The quality score cut-off (q) and percentage (p) value assign for pre-processing was 20 and 100, respectively (i.e. $q/20$ and $p/100$). After quality filtering we obtained a total paired-end (PE) reads of 75–192 million reads and a total singleton of 61–71 million reads with high-quality score (Table 7.1, Fig. 7.1).

De novo assembly of neem genome was performed on quality-filtered Illumina reads using Velvet program (Zerbino and Birney 2008). Also, we optimised the k-mer size for Velvet by iteratively testing various k-mers (27–67 nts) and comparing the assembly statistics. The selected k-mer size such as 45, 45 and 33 were emerged as the best choice for performing assembly of neem Genotypes 1, 2 and 3, respectively (Kuravadi et al. 2015). The parameter used for deciding the best k-mer were theoretical coverage, N50, maximum contigs length, totals contigs, assembled genome size and total number of reads used (Table 7.2).

© Springer Nature Switzerland AG 2019
M. Gowda et al. (eds.), *The Neem Genome*, Compendium of Plant Genomes,
https://doi.org/10.1007/978-3-030-16122-4_7

Table 7.1 Raw data filtering statistics

	A. indica 1	A. indica 2	A. indica 3
Total number of paired-end reads from Illumina	344,092,992	358,554,134	439,597,872
Number of paired-end reads obtained after pre-processed	118,014,568	192,086,722	75,284,296
Number of unpaired reads obtained after pre-processed	61,673,529	46,452,901	3,565,579
Total number of reads from Roche 454 GS-FLX+	3,008,843	–	–

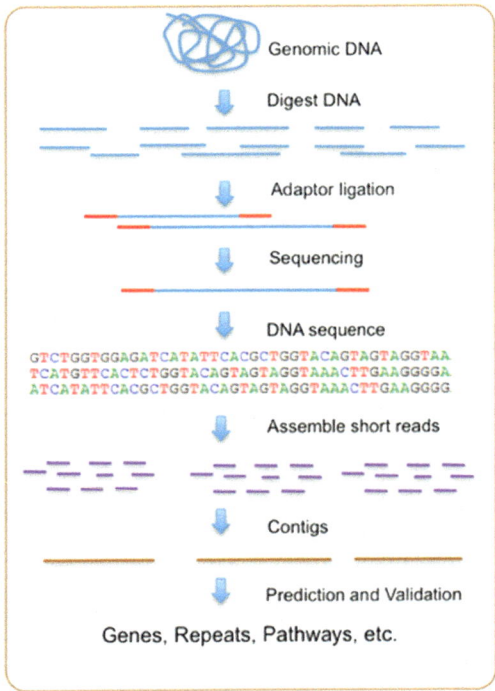

Fig. 7.1 Workflow of de novo shotgun genome sequencing and analysis

7.2 Neem Genome Assembly Using 454 Reads

The 454 long reads from Genotype 1 were assembled using MIRA assembler (which is a whole genome shotgun and EST sequence assembler) to obtain scaffolds from reads. The neem assembly statistics for MIRA assembly are summarised in Table 7.3.

7.3 Hybrid Assembly Using Illumina Contigs and 454 Reads

Hybrid assembly was carried out by combining Illumina and Roche 454 reads using a clustering approach. The Genotype 1 assembled contigs obtained from Velvet (Zerbino and Birney 2008) and MIRA (Chevreux 2005) assembler was merged using clustering program CD-HIT-est (Li and Godzik 2006), keeping minimum similarity cut-off of 90%. This clustering approach allowed both Illumina contigs and 454 reads to merge and

Table 7.2 Summary of whole genome assembly

	Genotype 1	Genotype 2	Genotype 3
k-mer used	45	45	33
Assembled genome size (Mb)	~216	~219	~213
Total number of contigs	94,780	87,734	175,452
Max. contigs lengths (bp)	241,126	250,005	135,514
N50 (bp)	22,263	32,639	16,213
Percentage bases in contigs greater than 1000 bp (%)	93.28	94.13	90.06
Total transposable element (%)	27.41	27.63	26.76

7 Neem Genome Assembly

Table 7.3 Assembly statistics for MIRA assembly of Roche 454 reads

	MIRA assembly statistics
Total length of sequence	~157 Mb
Total number of sequences	121,184
Minimum length of sequence	52 bp
Maximum length of sequence	43,859 bp
N50 stats	1463 bp
% of bases in contigs ≥ 1000 bp	74.54

Table 7.4 Assembly statistics for hybrid assembly and Illumina data assembly

	Hybrid assembly	Velvet assembly
Total length of sequence (Mb)	~267	~218
Total number of sequences	68,604	94,780
Minimum length of sequence (bp)	89	89
Maximum length of sequence (bp)	241,170	241,170
Average length of sequeces (bp)	3904.80	2302.56
N50 stats (bp)	15,948	22,041
GC (%)	31.38	31.13
N (%)	1.81	2.22

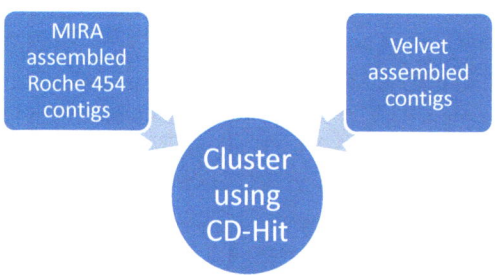

Fig. 7.2 The schematic diagram to shown the assembly strategy for hybrid assembly

build longer contigs of the genome. A total of 94,780 Illumina PE assembled contigs were clustered with a total of 121,184 Roche 454 reads giving a total of 68,604 unique sequences (Table 7.4, Fig. 7.2).

7.4 Evaluating Neem Genome Assembly

The completeness of the assembled genome was checked by using eukaryotic core gene-mapping approach (CEGMA) (Parra et al. 2007). CEGMA uses 248 core eukaryotic genes that are highly conserved and single-copy number in all types of eukaryotic genomes. Higher coverage of core genes shows better assembly of the genome. The analysis of neem hybrid assembly showed that 224 (90.32%) genes out of 248 core genes were present.

7.5 Transcriptome Assembly Using Trinity Assembler

Trinity efficiently constructs and analyses de Bruijn graphs, Trinity fully reconstructs a large fraction of transcripts, including alternatively spliced isoforms and transcripts from recently duplicated genes. Trinity recovers more full-length transcripts across a broad range of expression levels, with a sensitivity similar to methods that rely on genome alignments. This approach provides a unified solution for transcriptome reconstruction in any sample, especially in the absence of a reference genome (Grabherr et al. 2011; Henschel et al. 2012) (Fig. 7.3).

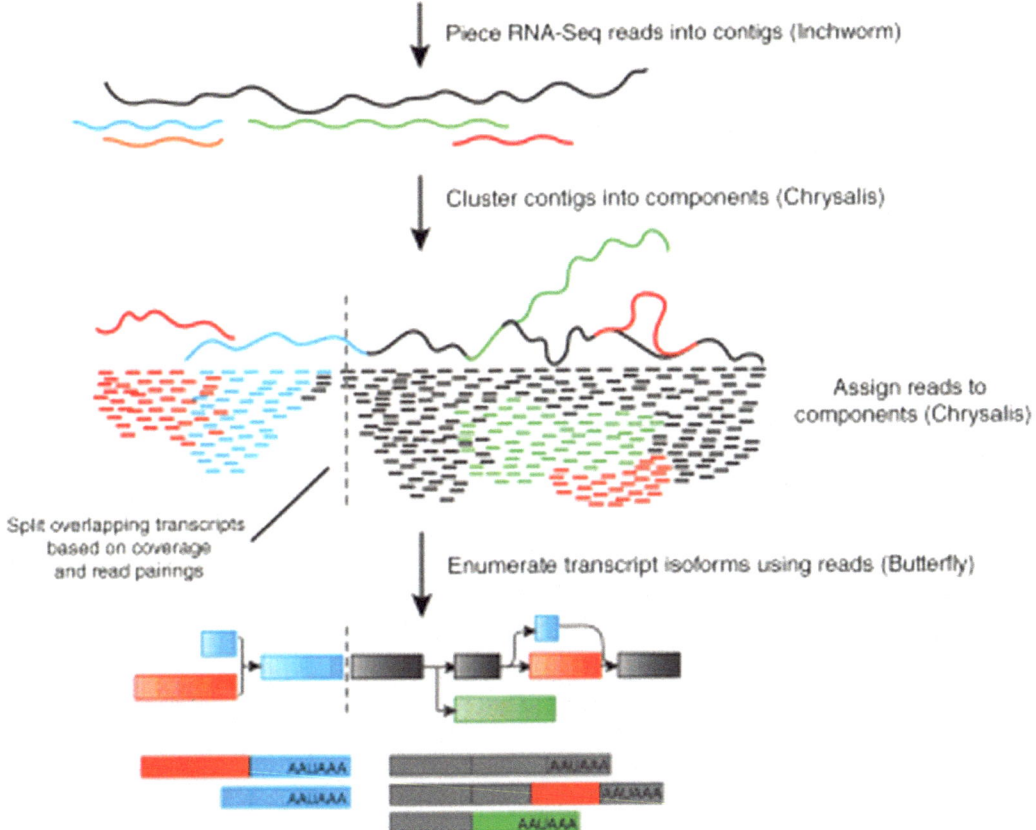

Fig. 7.3 Schematic representation of transcriptome assembly process using trinity (http://evomics.org/learning/genomics/trinity/)

7.6 Chloroplast and Mitochondrial Genome Assembly

Apart from assembling neem genome we also extracted reads corresponding to cell organelles such as chloroplast and mitochondria to generate a separate assembly for them. The reads for chloroplast and mitochondria were extracted separately by mapping genome reads to chloroplast and mitochondrial genomes of known plants; *A. thaliana*, *B. napus*, *C. papaya*, *C. sinensis*, *N. tabacum*, *P. dactylifera*, *P. trichocarpa*, *R. communis*, *S. bicolour* and *V. vinifera* using Bowtie2 (Langmead and Salzberg 2012). The mapped reads were extracted using Samtools (Li et al. 2009) and assembled separately using Velvet (Zerbino and Birney 2008).

The chloroplast genome assembly contains 60 scaffolds with size of 112,958 bp, which accounted for 72% of plant's chloroplast genome. The chloroplast genome had GC content of 38.07% with N50 of 2125 nts. Longest scaffold length was 8435 nts. The mitochondrial genome of neem was also assembled separately. The sequence covered 266,430 bp of the genome in 348 scaffolds. The GC content of the mitochondrial genome was 43.10%. The assembly statistics for organelle genomes has been summarised in Table 7.5.

Table 7.5 Organelle genomes assembly statistics for neem Genotype 1

	Mitochondria	Chloroplast
Total high-quality reads used	15,659,391	22,211,576
k-mer used (nt)	63	61
Assembled genome size (bp)	266,430	112,958
Total number of contigs	348	152
N50 (bp)	1490	2125
Maximum contig length (bp)	9110	8435
Minimum contig length (bp)	125	121
% of bases in contigs ≥ 1000 bp	60.71	63.33
GC content (%)	43.1	38.07
No of genes predicted	39	77

7.7 Conclusion

The hybrid assembly of short reads and long reads provided a good genome assembly having a completeness of >90% based of CEGMA analysis. For better understanding of expressed genes in the genome, an assembly of transcripts were generated using RNA from different tissues of neem tree. Also a separate assembly and annotation was done for organelle genomes that cover most of the organelle genes. The dataset with cellular genome, organelle genome and transcriptome provided detailed architecture of the need genome.

Acknowledgements We acknowledge Genomics facility (BT/PR3481/INF/22/140/2011) at Centre for Cellular and Molecular Platforms, Bangalore for sequencing of Neem genomes. We acknowledge Pradeep H, Aarati Karaba, Manojkumar S and Annapurna for their help in NGS library preparation and sequencing. We thank Ashmita G and Divya S for their help in manual curation of SSR markers. We are grateful to Rajanna, National Botanical Garden, University of Agricultural Sciences, GKVK campus, Bangalore for his help during neem sample collection.

References

Chevreux B (2005) MIRA: an automated genome and EST assembler. Ruprecht-Karls University, Heidelberg, Germany

Grabherr MG, Haas BJ, Yassour M, Levin JZ, Thompson DA, Amit I, Adiconis X, Fan L, Raychowdhury R, Zeng Q (2011) Full-length transcriptome assembly from RNA-Seq data without a reference genome. Nat Biotechnol 29:644–652

Henschel R, Lieber M, Wu LS, Nista PM, Haas BJ, LeDuc RD (2012) Trinity RNA-Seq assembler performance optimization. In: Proceedings of the 1st conference of the extreme science and engineering discovery environment: bridging from the eXtreme to the campus and beyond, p 45

Kuravadi NA, Yenagi V, Rangiah K, Mahesh HB, Rajamani A, Shirke MD, Russiachand H, Loganathan RM, Lingu CS, Siddappa S (2015) Comprehensive analyses of genomes, transcriptomes and metabolites of neem tree. PeerJ 3:e1066

Langmead B, Salzberg SL (2012) Fast gapped-read alignment with Bowtie 2. Nat Methods 9:357–359

Li W, Godzik A (2006) Cd-hit: a fast program for clustering and comparing large sets of protein or nucleotide sequences. Bioinformatics 22:1658–1659

Li H, Handsaker B, Wysoker A, Fennell T, Ruan J, Homer N, Marth G, Abecasis G, Durbin R, S. Genome Project Data Processing (2009) The sequence alignment/map format and SAMtools. Bioinformatics 25:2078–2079

Parra G, Bradnam K, Korf I (2007) CEGMA: a pipeline to accurately annotate core genes in eukaryotic genomes. Bioinformatics 23(9):1061–1067

Zerbino DR, Birney E (2008) Velvet: algorithms for de novo short read assembly using de Bruijn graphs. Genome Res 18:821–829

Repetitive Sequences

Nagesh A. Kuravadi and Malali Gowda

Abstract

Repetitive DNA sequences are an important component of eukaryotic genome. With the availability of the neem genome, it was used for identifying and classifying repetitive elements of the genome. This chapter details the analysis of transposable elements in neem that revealed the presence of 87 MB of reparative sequences in repeat families like SINEs, LINEs, and LTRs. Analysis of simple sequence repeats (SSR) was also performed that identified thousands of SSRs in each genotype. Further in silico polymorphism identification and filtration was done to shortlist polymorphic SSR's among the neem genotypes that can be used as genetic markers in diversity studies and mapping genes.

N. A. Kuravadi
Centre for Cellular and Molecular Platforms,
National Centre for Biological Sciences, Bengaluru,
Karnataka, India
e-mail: alwaysnagesh@gmail.com

M. Gowda (✉)
Center for Functional Genomics and Bio-Informatics, The University of TransDisciplinary and Health Sciences, Bengaluru, India
e-mail: malalig@tdu.edu.in

8.1 Introduction

Repetitive DNA sequence motifs repeated hundreds or thousands of times in the genome is found in most of the eukaryotic genomes (Biscotti et al. 2015). However, the significance of repetitive DNA in the genome is not completely understood, and it has been considered to have both structural and functional roles. High-throughput DNA sequencing reveals huge numbers of repetitive sequences. The prominent kind of repeat elements can be classified as Transposable elements (TE) and microsatellite DNA. In this chapter, we will discuss on the process of identification and analysis of repeat elements in neem genome.

8.2 Repeat Identification and Characterization

Transposable elements (TE) are the major components in eukaryotic genomes and contributed about 1000-fold genome size variation in plants (Zhang and Wessler 2004). De novo repeat identification was done using Repeat Modeler (http://www.repeatmasker.org/RepeatModeler.html). The program was run with RM-BLAST (NCBI) database as an input for repeat modelling. TE predicted in neem genome is about 32.53%, of which 17.06% is not annotated to any known repeat families. The long terminal repeat (LTR) retro-transposons are the major class of known repeats which accounted for 10% of repeat content in the neem genome. The details of repeats predicted in *A. indica* genome are tabulated in Table 8.1. We obtained 87 Mb (32.53%) of repeats in the Genotype 1 of neem genome (hybrid assembly) based on the de novo repeat prediction method. Our repeat prediction in neem (32.53%) is comparable to other plants genomes. The major dominant known repeat families like SINEs, LINEs, LTRs and DNA elements were accounted for 13.56% of neem genome in our study. We have also predicted 2.52% of low complexity and simple sequence repeats.

8.3 DNA Marker Identification

Simple sequence repeats (SSR), single-nucleotide polymorphisms (SNPs) and small insertions and deletions (InDels) are most abundant DNA markers in plant genomes. They are used to characterize plant genotypes and to differentiate between different sample sources (Godoy and Jordano 2001). DNA markers act as important tools in varietal characterization and also breeding. The SSRs and InDels are the efficient molecular markers because of their co-dominant inheritance, high abundance, enormous extent of allelic diversity and the ease of assessing SSR size variation by PCR with a pair of flanking primers (Kuravadi et al. 2014).

8.3.1 SSR Identification

Identification of simple sequence repeats (SSRs) or microsatellite was done using MIcroSAtellite tool (MISA) (http://pgrc.ipk-gatersleben.de/misa/). We identified 140,807, 108,020, 95,840 SSRs from Genotype 1, 2 and 3, respectively. SSRs motifs were also found in genes (2217, 1665 and 1606 genes in neem Genotypes 1, 2 and 3, respectively). We observed that corresponding SSR unit size in three neem genomes are quite similar. However, tri-repeats are highest (1841 SSRs) for genes as compared to other repeat units. AAG/CTT (533) repeat was the most prevalent groups of SSRs in genes as followed by mono- and di-repeats (Tables 8.2 and 8.3). Most prevalent mono-repeats was A/T (127) and di-repeat was AG/CT (114).

With the availability of three neem genomes, we used in-house in silico SSR analysis pipeline to identify polymorphic SSRs among neem genotypes (Fig. 8.1) (Kuravadi et al. 2015). In brief, the following steps were performed; The SSRs containing contigs were extracted for sequence-based SSRs repeat variability prediction. The identified SSR region along with 100 bp upstream and downstream from each SSR loci was extracted. The SSR regions were aligned to each other for a pair of genome using Bowtie2 alignment. The resulting SAM files were Parsed using the libraries functions of genomic ranges (Aboyoun et al. 2010), Gtools

Table 8.1 Details of repeats predicted in *A. indica* genome hybrid assembly

Repeat type	Subclass	Number of elements	Length (bp) occupied	Percentage of sequences
LINEs		2554	1,240,701	0.46
	LINE1	1918	1,002,661	0.37
	LINE2	119	39,194	0.01
LTR elements		51,260	26,838,058	10.02
	ERV_class1	462	130,859	0.05
DNA elements		17,356	7,112,032	2.65
Unclassified		166,584	45,693,338	17.06
Total interspersed repeats			80,894,162	30.20
Simple repeats		43,430	1,601,548	0.60
Low complexity		99,976	5,150,192	1.92

8 Repetitive Sequences

Table 8.2 Simple sequence repeats (SSRs) prediction from neem genotypes

	Genotype 1	Genotype 2	Genotype 3
Total number of sequences examined	68,604	87,734	175,452
Total size of examined contigs (bp)	267,885,489	221,053,547	215,992,204
Total number of identified SSRs	140,807	108,020	95,840
Number of SSR containing contigs	36,080	13,512	20,744
Number of contigs containing more than 1 SSR	20,182	8897	12,525
Number of SSRs present in compound formation	13,052	8908	7620

Table 8.3 Distribution of SSR family in genomes from neem genotypes

Unit size	Genotype 1 Number of SSRs	Genotype 2 Number of SSRs	Genotype 3 Number of SSRs
Mono-repeats	77,973	59,823	50,213
Di-repeats	42,942	32,899	30,363
Tri-repeats	13,289	10,201	10,188
Tetra-repeats	4665	3636	3641
Penta-repeats	1344	1024	1008
Hexa-repeats	594	437	427

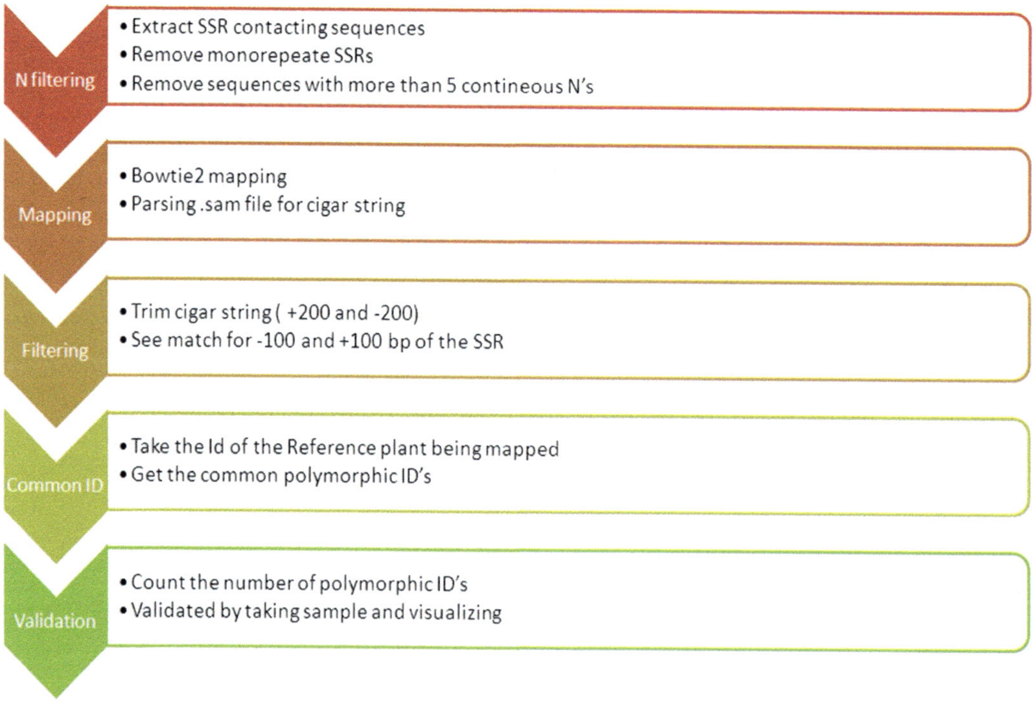

N filtering
- Extract SSR contacting sequences
- Remove monorepeate SSRs
- Remove sequences with more than 5 continous N's

Mapping
- Bowtie2 mapping
- Parsing .sam file for cigar string

Filtering
- Trim cigar string (+200 and -200)
- See match for -100 and +100 bp of the SSR

Common ID
- Take the Id of the Reference plant being mapped
- Get the common polymorphic ID's

Validation
- Count the number of polymorphic ID's
- Validated by taking sample and visualizing

Fig. 8.1 Overview of polymorphic SSR pipeline

(Warnes et al. 2008), Stringr (Wickham 2010) in R program to get list of polymorphic SSR regions between genomes. The concordance was also taken with neem tree Genotype 1 as reference to shortlist the most polymorphic SSRs. Number of polymorphic SSR's identified in comparison to Genotype 1 has been summarized in Table 8.4 (Fig. 8.2).

Table 8.4 Highly polymorphic di (p2), tri (p3) and tetra (p4) SSR motif from neem genome

SSR ID	SSR type	SSR	Size
G1_V2_SSR_54894	p2	(AC)18	36
G1_V2_SSR_18360	p2	(TC)16	32
G1_V2_SSR_8521	p2	(CA)15	30
G1_V2_SSR_16092	p2	(GA)14	28
G1_V2_SSR_11651	p2	(TA)14	28
G1_V2_SSR_8295	p2	(AC)13	26
G1_V2_SSR_45205	p2	(TA)13	26
G1_V2_SSR_61141	p2	(CT)13	26
G1_V2_SSR_62619	p2	(TA)12	24
G1_V2_SSR_5411	p2	(TC)12	24
G1_V2_SSR_11201	p2	(AT)12	24
G1_V2_SSR_56316	p2	(AG)12	24
G1_V2_SSR_23635	p2	(AT)12	24
G1_V2_SSR_13369	p2	(AT)12	24
G1_V2_SSR_8849	p2	(AT)12	24
G1_V2_SSR_47279	p2	(TA)11	22
G1_V2_SSR_15779	p2	(TA)11	22
G1_V2_SSR_3103	p2	(AT)11	22
G1_V2_SSR_30082	p2	(TA)11	22
G1_V2_SSR_35463	p2	(AT)11	22
G1_V2_SSR_16885	p2	(AT)11	22
G1_V2_SSR_8118	p2	(GA)11	22
G1_V2_SSR_58041	p2	(AT)11	22
G1_V2_SSR_62394	p2	(AT)11	22
G1_V2_SSR_55814	p2	(AG)11	22
G1_V2_SSR_10515	p2	(TA)11	22
G1_V2_SSR_41019	p2	(AT)11	22
G1_V2_SSR_35243	p2	(AT)11	22
G1_V2_SSR_32888	p2	(AT)11	22
G1_V2_SSR_52900	p2	(TA)11	22
G1_V2_SSR_4256	p2	(CT)11	22
G1_V2_SSR_32486	p2	(GA)11	22
G1_V2_SSR_32573	p2	(AT)11	22
G1_V2_SSR_54139	p2	(CT)11	22

(continued)

8 Repetitive Sequences

Table 8.4 (continued)

SSR ID	SSR type	SSR	Size
G1_V2_SSR_20577	p2	(TG)11	22
G1_V2_SSR_55360	p2	(AT)11	22
G1_V2_SSR_52588	p2	(AT)11	22
G1_V2_SSR_47732	p2	(TA)11	22
G1_V2_SSR_53170	p3	(TAT)16	48
G1_V2_SSR_37317	p3	(AAT)15	45
G1_V2_SSR_51471	p3	(TAT)15	45
G1_V2_SSR_3850	p3	(ATT)14	42
G1_V2_SSR_23087	p3	(GAT)14	42
G1_V2_SSR_485	p3	(GAT)14	42
G1_V2_SSR_3701	p3	(AAT)13	39
G1_V2_SSR_34277	p3	(TTA)13	39
G1_V2_SSR_62729	p3	(TAA)13	39
G1_V2_SSR_34775	p3	(GAA)12	36
G1_V2_SSR_35585	p3	(TTA)12	36
G1_V2_SSR_19633	p3	(ATT)12	36
G1_V2_SSR_31573	p3	(TAA)12	36
G1_V2_SSR_26487	p3	(AAT)11	33
G1_V2_SSR_27294	p3	(TAA)11	33
G1_V2_SSR_23015	p3	(ATA)11	33
G1_V2_SSR_10239	p3	(TTA)11	33
G1_V2_SSR_8051	p3	(TAA)10	30
G1_V2_SSR_46433	p3	(GAA)9	27
G1_V2_SSR_11475	p3	(ATT)9	27
G1_V2_SSR_33796	p3	(TTA)9	27
G1_V2_SSR_34185	p3	(AAG)9	27
G1_V2_SSR_2688	p3	(ATA)9	27
G1_V2_SSR_34912	p3	(ATC)9	27
G1_V2_SSR_20455	p3	(TTA)9	27
G1_V2_SSR_27104	p3	(ATC)8	24
G1_V2_SSR_11771	p3	(AAT)8	24
G1_V2_SSR_26005	p3	(ATT)8	24
G1_V2_SSR_8856	p3	(TGC)8	24
G1_V2_SSR_28583	p3	(CAT)8	24
G1_V2_SSR_11536	p3	(AAC)8	24
G1_V2_SSR_24254	p3	(GTT)7	21
G1_V2_SSR_32847	p3	(TCT)7	21
G1_V2_SSR_4319	p4	(TACA)5	20
G1_V2_SSR_25974	p4	(ACAT)9	36

(continued)

Table 8.4 (continued)

SSR ID	SSR type	SSR	Size
G1_V2_SSR_28750	p4	(CACG)5	20
G1_V2_SSR_25853	p4	(TCAA)6	24
G1_V2_SSR_20810	p4	(AATA)6	24
G1_V2_SSR_63018	p4	(TAAT)7	28
G1_V2_SSR_21627	p4	(TGAT)6	24
G1_V2_SSR_28280	p4	(AACG)5	20
G1_V2_SSR_10350	p4	(AAAT)8	32
G1_V2_SSR_27817	p4	(AAAT)5	20
G1_V2_SSR_11531	p4	(AAAT)6	24
G1_V2_SSR_4373	p4	(TTAT)6	24
G1_V2_SSR_20225	p4	(TTAT)7	28
G1_V2_SSR_36199	p4	(ATTA)6	24
G1_V2_SSR_5487	p4	(TTAA)8	32

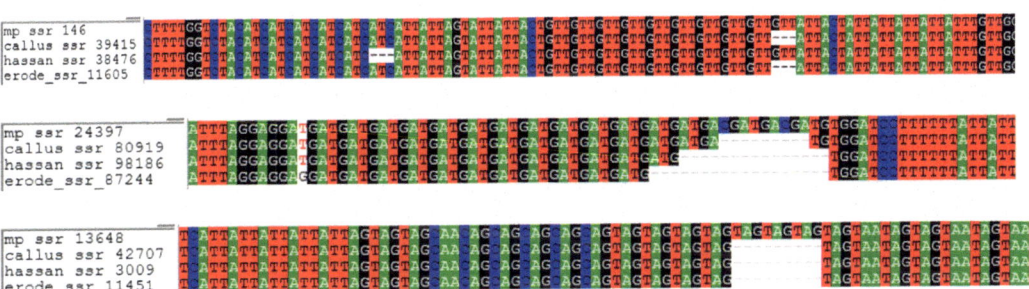

Fig. 8.2 Visualization of polymorphic SSR's sequences that were identified using polymorphic SSR identification pipeline

8.4 Conclusion

Repetitive elements form an important component of eukaryotic genome. Analysis of neem genome revealed 32.53% of repetitive elements comprising mainly of known repeat families like SINEs, LINEs, and LTRs. However, there were 17% of unclassified repeat types in neem genome. Analysis of microsatellites with comparison with other genotypes yielded SSR markers polymorphic among the lines.

Acknowledgements We acknowledge Genomics facility (BT/PR3481/INF/22/140/2011) at Centre for Cellular and Molecular Platforms, Bangalore for sequencing of Neem genomes. We acknowledge Pradeep H, Aarati Karaba, Manojkumar S and Annapurna for their help in NGS library preparation and sequencing. We thank Ashmita G and Divya S for their help in manual curation of SSR markers. We are grateful to Rajanna, National Botanical Garden, University of Agricultural Sciences, GKVK campus, Bangalore for his help during neem sample collection.

References

Aboyoun P, Pages H, Lawrence M (2010) GenomicRanges: representation and manipulation of genomic intervals. R Package Version 1:1–25

Biscotti MA, Olmo E, Heslop-Harrison JSP (2015) Repetitive DNA in eukaryotic genomes. Springer

Godoy JA, Jordano P (2001) Seed dispersal by animals: exact identification of source trees with endocarp DNA microsatellites. Mol Ecol 10:2275–2283

Kuravadi NA, Tiwari PB, Tanwar UK, Tripathi SK, Dhugga KS, Gill KS, Randhawa GS (2014) Identification and characterization of EST-SSR markers in cluster bean (*Cyamopsis* spp.). Crop Sci 54:1097–1102

Kuravadi NA, Yenagi V, Rangiah K, Mahesh HB, Rajamani A, Shirke MD, Russiachand H, Loganathan RM, Lingu CS, Siddappa S (2015) Comprehensive analyses of genomes, transcriptomes and metabolites of neem tree. PeerJ 3:e1066

Warnes GR, Bolker B, Lumley T (2008) gtools: Various R programming tools. R Package Version 2

Wickham H (2010) stringr: Make it easier to work with strings. R Package Version 0.4. URL http://CRAN.R-project.org/package=stringr

Zhang X, Wessler SR (2004) Genome-wide comparative analysis of the transposable elements in the related species *Arabidopsis thaliana* and *Brassica oleracea*. Proc Natl Acad Sci U S A 101:5589–5594

Neem Genome Annotation

Nagesh A. Kuravadi and Malali Gowda

Abstract

Gene annotation is considered an important step in understanding the functioning of the genome. With a good genome assembly in place, gene prediction and annotation is done to understand the functional aspects of the genome. However, the process becomes more complex due to the presence of alternative splicing and other regulatory factors that affect the protein-coding genes. To make the gene prediction more accurate, we used two different gene prediction programs to identify the genes. The genome annotation was carried out by mapping the gene sequences to multiple databases to obtain gene function. Along with annotation of nuclear genes, prediction and annotation of mitochondrial and chloroplast genes have been summarized in this chapter.

9.1 Introduction

Gene annotation is considered an important step after genome assembly of an organism. It provides information on the functional genes and pathways in the organism. The process can be divided into two parts as gene prediction and gene annotation. During gene prediction, the nucleotide bases of the assembled genome are screened with a gene model developed from well-characterized species like *Arabidopsis*. The process provides a set of sequences from the genome along with the information on the contigs from with they originate. However, gene prediction is considered complicated by the existence of transcriptional complexity, which includes alternative splicing and transcriptional events outside of protein-coding genes (Mudge and Harrow 2016). Hence, we can see the use of multiple gene prediction algorithms to predict genes and conclude them.

The process of annotation can be understood as defining the function of the predicted genes. This is done by comparing the genes predicted to gene databases with functional information like NCBI's Nr database, UniProt database, Gene Ontology, Kyoto Encyclopedia of Genes and Genomes (KEGG) and Enzyme Commission number (EC) databases. This comparison results with a list of information for each gene and characteristic functions that similar genes perform in other organisms resulting in a table of

N. A. Kuravadi
Centre for Cellular and Molecular Platforms,
National Centre for Biological Sciences, Bengaluru,
Karnataka, India
e-mail: alwaysnagesh@gmail.com

M. Gowda (✉)
Center for Functional Genomics and
Bio-Informatics, The University of
TransDisciplinary and Health Sciences,
Bengaluru, India
e-mail: malalig@tdu.edu.in

annotation information about the predicted functions for the gene.

9.2 Gene Prediction

Gene prediction is considered as a difficult process in a draft genome assembly due to lack of good gene prediction models. To overcome this difficulty, we used an approach of predicting genes using two different programs and combining their output to get complete set of genes. Genes were predicted on the assembled genome using Augustus (Stanke et al. 2006) and GenScan (Burge and Karlin 1997). The total number of genes predicted using two programs are varied. Due to the difference in genes prediction by the two program, we aligned two sets using BLAT (Kent 2002). Then we clustered genes with similarity cut-off of 90% using CD-HIT-est program (Li and Godzik 2006). This method gave the overall representation of genes by merging similar genes predicted by both the programs (Kuravadi et al. 2015). Then we discarded genes which are less than 100 bp (Delcher et al. 1999). Genes which qualified minimum threshold were mapped with RNA-Seq data to validate expression (Table 9.1).

9.3 Gene Annotation

We used all annotated neem genes with and without RNA-seq evidence to search function using UniProt database, Gene Ontology (GO), Kyoto Encyclopedia of Genes and Genomes (KEGG) and Enzyme Commission number (EC). Schematic representation of GO classes present in neem tree genome is summarized in Fig. 9.1. Below analysis was done using BLASTX (Altschul et al. 1990) in annot8r (Schmid and Blaxter, 2008) with E-value cut-off of 10^{-3}. Genes with multiple hit were filtered based on E-value using perl script.

9.4 Chloroplast and Mitochondrial Gene Prediction

Further, gene prediction of chloroplast genome was done using DOGMA (Wyman et al. 2004). The gene prediction showed 77 unique genes in the chloroplast genome of neem. Mitofy predicted 39 mitochondrial genes out of 41 genes in other plants. The annotation for the predicted genes was also obtained.

9.5 Gene Expression Analysis

The consolidated 44,495 genes predicted in neem genome, which were mapped with RNA-seq data from individual tissues using SeqMap (Jiang and Wong, 2008). The mapping data from individual tissues were used to measure the expression value in RPKM (reads per kilo base per million) using rSeq tool (Jiang and Wong, 2009). RPKM was chosen as it provides normalized expression value for the genes in the tissue being assessd. RPKM counts the number of reads mapped to a gene for

Table 9.1 Summary of gene prediction using Augustus and GeneScan

Number of predicted genes in Augustus	40,130
Number of predicted genes in Gene Scan	52,617
Number of Genes clustered from Gene Scan and Augustus	48,032
No of genes with >100 bp	44,495
Genes with RNA-seq evidence	32,278
Genes without transposable elements in them	29,050

Fig. 9.1 GO classification of neem genes

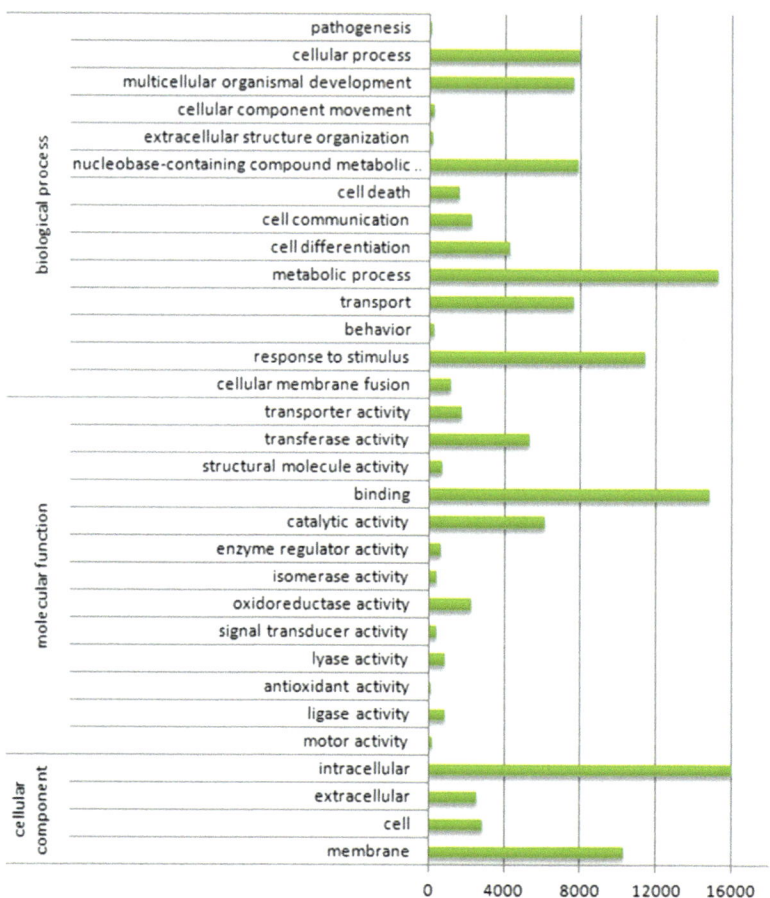

Table 9.2 Number of genes at different expression level in different tissues of neem

Tissue	Quality filtered reads (q20p100)	No of genes/RPKM >1	No of genes/RPKM >5	No of genes/RPKM >10
Mature Leaf	5,401,910	19,308	14,807	11,763
Flower and bud	22,654,982	21,927	16,716	13,632
Fruit coat and pulp	55,627,021	21,537	16,693	13,888
Developing endosperm	31,340,522	19,480	15,262	12,614
Mature fruit	23,321,657	17,366	12,407	9741

each million reads in the transcriptome. The genes with RPKM expression value 1 and above were used for further analysis (Davidson et al. 2011). The RPKM value was used to cluster the genes according to their expression pattern using WCGNA package in R (Langfelder and Horvath 2008). Table 9.2 shows the number of genes in each tissue at different expression levels.

Further, the gene that are expressed commonly and tissue-specific manner were analyzed. The analysis showed that 3008 genes showing tissue-specific expression in various tissues, while 14,249 genes were expressed in all the tissues (Fig. 9.2). Table 9.3 shows 100 highly expressed genes unique to neem with gene expression comparison among various tissues.

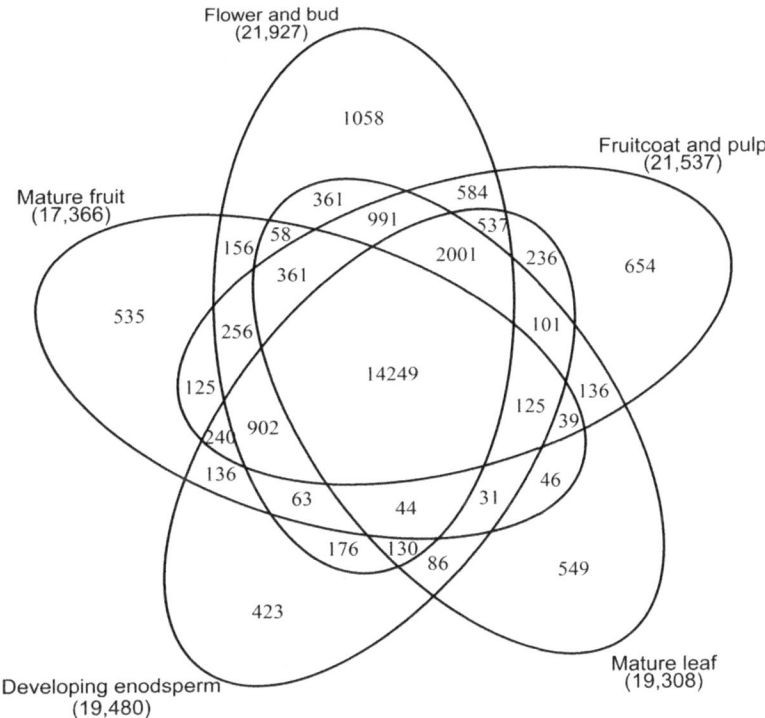

Fig. 9.2 Comparison of genes expressed commonly and uniquely in flower, mature endosperm, endosperm, leaf, fruit coat and pulp tissues

9.6 Gene Co-expression Analysis

To further understand the gene expression pattern of co-expression, we carried out co-expression analysis of all the neem genes in different tissues. It is a powerful, widely used method to investigate underlying patterns in gene expression data. This approach aims to find groups of genes with closely correlated expression profiles. Figure 9.3 shows the clustering of genes by co-expression.

The co-expression analysis showed that neem genes can be divided into 28 major groups as shown in Table 9.4.

9.7 Transcript Expression Analysis

The combined transcriptome of all tissues of neem tree from Genotype 1 were assembled using Trinity software (Grabherr et al. 2011). The transcripts were clustered to remove over-represented short fragments using CD-HIT-est program (Li and Godzik 2006) with a minimum similarity cut-off of 90%. The expression values were also determined for assembled transcriptome and used for verifying the expression of genes predicted by gene models.

9.8 Annotation of Mitochondrial and Chloroplast Genes

Gene prediction and annotation of the mitochondrial assembly was done using Mitofy (Alverson et al. 2010). Mitofy predicted 39 mitochondrial genes with annotation. The details of the mitochondrial gene prediction have been summarized in Table 9.5 Gene prediction and assembly of chloroplast genome was done using DOGMA (Wyman et al. 2004). The list of genes predicted in chloroplast genome and its annotation has been listed in Table 9.6.

9 Neem Genome Annotation

Table 9.3 Top 100 highly expressed genes unique to neem with BLAST annotation

Gene ID	BLAST description	GO description	KEGG description	EC description	Mature leaf	Flower and bud	Developing endosperm	Fruit coat and pulp	Mature fruit	Seedling root	Seedling shoot
Ai02g15517	hypothetical protein ARALYDRAFT_903435 [Arabidopsis lyrata subsp. lyrata]gb|EFH56176.1| hypothetical protein ARALYDRAFT_903435 [Arabidopsis lyrata subsp. lyrata]	–	–	–	3.51194	1.99916	27.0554	7.27648	113351	9.17194	6.02999
Ai02g7151	Uncharacterized protein TCM_001312 [Theobroma cacao]	–	–	–	784.029	622.947	8570.98	973.705	18.3393	716.989	828.134
Ai02g6731	PREDICTED: carnitinyl-CoA dehydratase-like [Vitis vinifera]	–	–	–	928.012	2121.12	2693.24	1055.31	1008.54	646.271	306.415
Ai02g10923	unknown protein [Jatropha curcas]gb|ADI67178.1| hypothetical protein [Jatropha curcas]gb|ADI67180.1| F9L1.21 protein [Jatropha curcas]	–	–	–	543.419	1582.8	1048.04	1382.69	429.83	1708.25	884.119
Ai02g13231	late embryogenesis abundant protein [Catharanthus roseus]	–	–	–	1.06538	0	722	20.4184	5340.03	0	0
Ai02g1225	putative a2 protein [Neofusicoccum parvum UCRNP2]	–	–	–	169.565	566.329	3856.55	730.206	4.84785	93.7784	171.945
Ai02g1766	PREDICTED: uncharacterized protein LOC100258053 [Vitis vinifera]	–	–	–	501.876	1037.88	642.824	691.621	253.805	1079.06	667.331
Ai02g9238	Uncharacterized protein TCM_001312 [Theobroma cacao]	–	–	–	19.681	26.568	3985.88	81.0115	0.329847	275.503	450.112
Ai02g17152	Late embryogenesis abundant protein D-29, putative [Theobroma cacao]	–	–	–	6.22148	0.467019	184.551	6.7427	2100.27	0	4.93027
Ai02g27279	hypothetical protein POPTR_0004s09260 g, partial [Populus trichocarpa]	–	–	–	129.687	217.222	95.6035	121.116	1432.57	87.9598	83.7512

(continued)

Table 9.3 (continued)

Gene ID	BLAST description	GO description	KEGG description	EC description	Mature leaf	Flower and bud	Developing endosperm	Fruit coat and pulp	Mature fruit	Seedling root	Seedling shoot
Ai02g26011	protein SAH7 [Arabidopsis thaliana]emb[CAB40579.1] SAH7 protein [Arabidopsis thaliana]gb[AAL69502.1] unknown protein [Arabidopsis thaliana]gb[AAM45117.1] unknown protein [Arabidopsis thaliana]gb[AAM62935.1] allergen-like protein BRSn20 [Arabidopsis thaliana]dbj[BAF00475.1] hypothetical protein [Arabidopsis thaliana]gb[AEE82668.1] protein SAH7 [Arabidopsis thaliana]	–	–	–	127.192	294.211	215.113	515.35	7.10562	516.723	160.151
Ai02g43595	PREDICTED: eukaryotic initiation factor 4A-III-B-like [Tursiops truncatus]	–	–	–	81.8364	259.546	262.689	267.526	192.017	370.461	168.615
Ai02g21703	PREDICTED: F-box protein SKIP23-like [Vitis vinifera]	"biological_process"	–	–	4.39702	381.472	53.5216	1095.51	7.90776	1.06001	3.13601
Ai02g23294	Uncharacterized protein TCM_012229 [Theobroma cacao]	–	–	–	300.671	211.304	0.92247	223.038	2.79952	445.942	279.158
Ai02g20561	predicted protein [Populus trichocarpa]gb[ERP62179.1] hypothetical protein POPTR_0004s14520 g [Populus trichocarpa]	–	–	–	0	0.396626	2.72713	1436.23	1.43048	1.27378	2.51229
Ai02g3911	Arginine/serine-rich splicing factor 35 [Theobroma cacao]	–	–	–	64.8386	52.7274	200.315	258.097	37.5395	317.119	435.19
Ai02g16429	PREDICTED: major allergen Pru ar 1 [Vitis vinifera]emb[CBI22941.3] unnamed protein product [Vitis vinifera]	"response to biotic stimulus"	–	–	852.394	239.132	0	40.2813	0.544567	3.39438	198.612
Ai02g8683	hypothetical protein [Brassica napus]	–	–	–	106.635	117.462	134.307	141.603	2.43705	486.098	169.777
Ai02g28177	Uncharacterized protein TCM_001312 [Theobroma cacao]	–	–	–	20.7057	7.2216	820.01	35.62	1.94002	157.007	50.666

(continued)

9 Neem Genome Annotation

Table 9.3 (continued)

Gene ID	BLAST description	GO description	KEGG description	EC description	Mature leaf	Flower and bud	Developing endosperm	Fruit coat and pulp	Mature fruit	Seedling root	Seedling shoot
Ai02g31755	chitinase CHI1 [Citrus sinensis]	"metabolic process"	"Amino sugar and nucleotide sugar metabolism"	"Chitinase."	0	8.96509	0	0.152277	1043.92	0	0
Ai02g7243	Uncharacterized protein TCM_029557 [Theobroma cacao]	–	–	–	56.3281	127.544	121.197	135.282	261.772	186.369	104.996
Ai02g9344	Uncharacterized protein TCM_001312 [Theobroma cacao]	–	–	–	97.7491	156.15	551.305	22.3337	0	8.39296	96.5624
Ai02g4898	uncharacterized protein LOC100306298 [Glycine max] gb|ACU14415.1| unknown [Glycine max]	"plasma membrane"	–	–	164.248	3.54604	246.152	12.5308	11.556	0	472.208
Ai02g2583	PREDICTED: cytochrome b5-like [Hydra magnipapillata]	"mitochondrial inner membrane"	–	"Nitratereductase (NADH)"	40.071	8.88713	628.717	26.92	1.22105	60.888	63.798
Ai02g1552	Uncharacterized protein TCM_026440 [Theobroma cacao]	–	–	–	389.379	70.4317	1.28115	71.2097	0	60.8251	219.522
Ai02g29223	hypothetical protein PRUPE_ppa013918 mg [Prunus persica]	–	–	–	228.386	41.4539	0	5.4891	0	123.11	359.511
Ai02g18671	phytosulfokines precursor, putative [Ricinus communis] gb|EEF37131.1| phytosulfokines precursor, putative [Ricinus communis]	–	–	–	154.406	63.7949	28.7787	174.407	154.641	66.3418	115.453
Ai02g4135	conserved hypothetical protein [Ricinus communis] gb|EEF33214.1| conserved hypothetical protein [Ricinus communis]	"response to external stimulus"	"Plant hormone signal transduction"	–	17.7726	247.283	34.3004	422.994	6.05332	6.28856	6.49683
Ai02g7773	Disease resistance response protein, putative [Ricinus communis] gb|EEF43091.1| Disease resistance response protein, putative [Ricinus communis]	–	–	–	91.7049	21.559	452.634	46.6378	0	53.561	73.0027

(continued)

Table 9.3 (continued)

Gene ID	BLAST description	GO description	KEGG desciption	EC description	Mature leaf	Flower and bud	Developing endosperm	Fruit coat and pulp	Mature fruit	Seedling root	Seedling shoot
Ai02g22163	predicted protein [Populus trichocarpa]	"transition metal ion transport"	–	–	244.877	118.252	2.06605	94.1654	0.253336	142.118	99.6627
Ai02g9053	RNA binding protein, putative [Ricinus communis|gb|EEF50247.1] RNA binding protein, putative [Ricinus communis]	–	–	–	77.7731	126.058	135.425	99.5536	53.0307	93.4888	82.3168
Ai02g6378	hypothetical protein PRUPE_ppa013983 mg [Prunus persica]	–	–	–	153.606	58.8878	0.371812	36.5792	0	212.564	189.071
Ai02g8563	predicted protein [Populus trichocarpa]	"biological_process"	–	–	2.0674	198.847	5.34642	408.698	1.62416	25.1068	7.18816
Ai02g1574	predicted protein [Populus trichocarpa]	–	–	–	66.5526	62.6622	184.514	100.833	69.7994	84.7331	76.0613
Ai02g7499	hypothetical protein PRUPE_ppa012081 mg [Prunus persica]	–	–	–	3.01928	37.5662	76.8143	12.1986	509.688	0	1.55523
Ai02g7240	JHL20I20.12 [Jatropha curcas]	–	–	–	51.0186	74.4653	56.8901	87.4815	203.223	87.5265	72.0424
Ai02g1733	hypothetical protein RCOM_0452240 [Ricinus communis|gb|EEF30830.1] hypothetical protein RCOM_0452240 [Ricinus communis]	–	–	–	74.7852	59.4939	244.533	91.7064	35.762	67.5101	48.5709
Ai02g24469	proteophosphoglycan ppg4 [Leishmania braziliensis MHOM/BR/75/M2904|emb|CAM43270.1] proteophosphoglycan ppg4 [Leishmania braziliensis MHOM/BR/75/M2904]	"pseudohyphal growth"	–	–	74.2817	59.3929	62.9182	97.8516	122.34	110.818	73.3799
Ai02g17643	hypothetical protein POPTR_0016s08050 g [Populus trichocarpa]	"carbonate dehydratase activity"	"Nitrogen metabolism"	"Monodehydroascorbatereductase(NADH)"	254.051	117.347	0.180384	3.20838	2.55468	69.003	131.11
Ai02g23897	PREDICTED: 60S ribosomal protein L6-1-like [Cucumis sativus]	"structural constituent of ribosome"	"Ribosome"	–	11.014	68.5186	101.101	72.7392	141.904	116.497	59.5695

(continued)

Table 9.3 (continued)

Gene ID	BLAST description	GO description	KEGG description	EC description	Mature leaf	Flower and bud	Developing endosperm	Fruit coat and pulp	Mature fruit	Seedling root	Seedling shoot
Ai02g9546	Histidine kinase [Medicago truncatula]gb|AET03757.1| Histidine kinase [Medicago truncatula]	"intracellular"	–	"Histidinekinase"	0	172.148	5.13947	352.026	0	0	0
Ai02g24372	Uncharacterized protein isoform 1 [Theobroma cacao] gb|EOY22003.1| Uncharacterized protein isoform 1 [Theobroma cacao]	–	–	–	41.0581	49.2325	186.186	73.0366	16.7501	88.9536	51.9141
Ai02g38228	Nuclear fusion defective 6 isoform 2 [Theobroma cacao]	–	–	–	52.5219	64.067	88.1027	71.9431	51.3478	36.5783	132.264
Ai02g5940	hypothetical protein POPTR_0005s16202 g, partial [Populus trichocarpa]	–	–	–	22.0724	77.6955	88.1554	102.898	84.8185	59.2922	51.9746
Ai02g28752	vitamin-b12 independent methionine synthase [Populus trichocarpa]	"cytoplasm"	"Selenocompound metabolism"	"5-methyltetrahydropteroyltriglutamate–homocysteineS-methyltransferase"	28.4704	25.1153	33.6104	61.8895	58.6866	204.485	74.2642
Ai02g36432	hypothetical protein PRUPE_ppa016770 mg [Prunus persica]	"monooxygenase activity"	–	–	45.553	29.6353	268.152	97.9217	0.954314	11.8968	27.375
Ai02g25125	Calcium-binding EF-hand family protein [Theobroma cacao]	"biological_process"	"Plant-pathogen interaction"	–	17.1681	70.774	36.6853	320.576	5.36041	0	29.2344
Ai02g16395	GA-stimulated transcript-like protein 1 [Gossypium hirsutum]	"extracellular region"	–	–	5.17876	416.298	1.44785	30.8488	0.650956	12.1726	8.00272
Ai02g32427	Uncharacterized protein TCM_029860 [Theobroma cacao]	–	–	–	44.1812	39.3948	81.9309	105.612	78.2126	85.3063	27.3087
Ai02g4927	hypothetical protein POPTR_0006s12830 g [Populus trichocarpa]	–	–	–	0	0	0	8.51756	0	443.768	3.76452
Ai02g32666	hypothetical protein POPTR_0005s02820 g [Populus trichocarpa]	"transferase activity"	"Phenylpropanoid biosynthesis"	"AnthranilateN-benzoyltransferase"	248.683	8.72768	0.808849	4.40557	3.8366	11.143	162.643

(continued)

Table 9.3 (continued)

Gene ID	BLAST description	GO description	KEGG description	EC description	Mature leaf	Flower and bud	Developing endosperm	Fruit coat and pulp	Mature fruit	Seedling root	Seedling shoot
Ai02g8565	hypothetical protein POPTR_0001s40530 g [Populus trichocarpa]	"biological_process"	–	–	6.9459	49.2123	132.912	195.633	1.09135	30.3849	19.3799
Ai02g3471	PREDICTED: epoxide hydrolase 2-like [Glycine max]	–	–	–	26.0618	45.8115	15.2303	15.089	254.127	25.8552	45.6004
Ai02g41875	PREDICTED: uncharacterized protein LOC100255613 [Vitis vinifera]emb|CBI37757.3| unnamed protein product [Vitis vinifera]	–	–	–	33.1353	67.9037	68.552	45.6548	39.1511	67.4993	102.408
Ai02g1057	PREDICTED: uncharacterized protein LOC100241532 [Vitis vinifera]	–	–	–	21.2736	175.377	9.90507	145.142	3.63676	30.7446	15.8658
Ai02g8959	Histidine-containing phosphotransfer protein, putative [Theobroma cacao]	"cytoplasm"	"Plant hormone signal transduction"	–	31.9812	97.9054	58.4539	100.356	3.37157	48.4976	54.2031
Ai02g22985	conserved hypothetical protein [Ricinus communis]gb|EEF52875.1| conserved hypothetical protein [Ricinus communis]	–	–	–	30.4368	39.412	37.3103	59.1821	101.826	67.8724	57.8879
Ai02g1846	predicted protein [Populus trichocarpa]	–	–	–	56.1451	70.8516	43.1378	78.2079	44.5777	49.3693	41.1958
Ai02g15390	putative inclusion body protein [Sweet potato caulimo-like virus]gb|AEA36697.1| putative inclusion body protein [Sweet potato caulimo-like virus]	–	–	–	298.193	8.7754	0	1.04339	0.180854	0	55.2141
Ai02g39051	Mediator of RNA polymerase II transcription subunit 19 isoform 3 [Theobroma cacao]	–	–	–	6.48957	5.93705	84.3658	12.6055	41.6018	157.62	41.7846
Ai02g8400	Coiled-coil domain-containing protein, putative [Ricinus communis]gb|EEF38594.1| Coiled-coil domain-containing protein, putative [Ricinus communis]	–	–	–	30.4014	53.1539	45.4879	65.3516	63.6896	58.225	33.0595

(continued)

Table 9.3 (continued)

Gene ID	BLAST description	GO description	KEGG description	EC description	Mature leaf	Flower and bud	Developing endosperm	Fruit coat and pulp	Mature fruit	Seedling root	Seedling shoot
Ai02g14793	PREDICTED: alanine–glyoxylate aminotransferase 2 homolog 2, mitochondrial-like [Glycine max]	"catalytic activity"	"Alanine"	"Alanine–glyoxylatetransaminase"	7.9256	23.8471	0	13.6844	0	260.805	38.103
Ai02g16806	conserved hypothetical protein [Ricinus communis]gb|EEF32419.1| conserved hypothetical protein [Ricinus communis]	–	–	–	39.3951	17.7802	4.03842	13.5699	97.5516	9.25973	145.09
Ai02g9337	hypothetical protein POPTR_0017s12480 g [Populus trichocarpa]	–	–	–	43.5117	64.3457	24.7612	79.4079	45.409	27.6931	29.6327
Ai02g17043	MLP-like protein 423, putative [Theobroma cacao]	–	–	–	0	114.6	11.6054	186.972	0	0	0
Ai02g406	hypothetical protein POPTR_0009s08430 g [Populus trichocarpa]	–	–	–	75.607	27.9315	4.42887	7.60882	7.05982	27.08	162.457
Ai02g4260	predicted protein [Populus trichocarpa]	–	–	–	23.5247	41.5593	54.8831	50.8843	4.69039	75.6324	53.0664
Ai02g17478	Uncharacterized protein isoform 2 [Theobroma cacao]	–	–	–	120.307	80.2753	2.34271	43.0635	0.451409	23.9166	31.4473
Ai02g27666	PREDICTED: protein EARLY RESPONSIVE TO DEHYDRATION 15-like isoform 1 [Cucumis sativus]|ref|XP_004138407.1| PREDICTED: protein EARLY RESPONSIVE TO DEHYDRATION 15-like isoform 2 [Cucumis sativus]|ref|XP_004167178.1| PREDICTED: protein EARLY RESPONSIVE TO DEHYDRATION 15-like [Cucumis sativus]	–	–	–	14.218	44.6592	34.45	49.1593	7.14868	103.043	43.9422
Ai02g7722	conserved hypothetical protein [Ricinus communis]gb|EEF37664.1| conserved hypothetical protein [Ricinus communis]	–	–	–	12.2388	34.6983	10.6354	62.3167	142.97	25.2342	5.74133

(continued)

Table 9.3 (continued)

Gene ID	BLAST description	GO description	KEGG description	EC description	Mature leaf	Flower and bud	Developing endosperm	Fruit coat and pulp	Mature fruit	Seedling root	Seedling shoot
Ai02g6767	Uncharacterized protein TCM_016791 [Theobroma cacao]	–	–	–	23.2263	32.9966	20.6953	69.0692	34.4071	84.846	28.2175
Ai02g10347	conserved hypothetical protein [Ricinus communis]gb|EEF31175.1| conserved hypothetical protein [Ricinus communis]	–	–	–	14.1894	29.8089	32.1767	33.9726	65.1993	38.2929	77.9619
Ai02g10770	Organ-specific protein P4, putative [Theobroma cacao]	–	–	–	12.7039	90.2231	1.31543	18.2788	7.80677	143.771	17.45
Ai02g10396	F-box family protein, putative [Theobroma cacao]	"biological_process"	–	–	198.252	33.2996	2.27276	11.6953	4.8455	15.4096	23.9089
Ai02g15110	chromatin remodeling complex subunit [Populus trichocarpa] gb|EEE94860.1| MORPHEUS MOLECULE family protein [Populus trichocarpa]	"2 iron"	–	–	17.9281	15.9927	59.9173	30.5955	107.962	33.2399	22.9811
Ai02g16576	Phloem protein 2-A10, putative [Theobroma cacao]	"carbohydrate binding"	–	–	96.9991	30.0071	0	12.4314	0	54.844	89.6263
Ai02g22182	Uncharacterized protein TCM_036899 [Theobroma cacao]	–	–	–	39.8588	36.118	39.7983	49.649	13.3604	46.1001	55.7276
Ai02g595	Uncharacterized protein TCM_025735 [Theobroma cacao]	–	–	–	14.7237	32.5279	71.6554	37.7093	28.7891	52.5525	29.494
Ai02g19269	Dynein light chain type 1 family protein isoform 1 [Theobroma cacao]gb|EOY28429.1| Dynein light chain type 1 family protein isoform 1 [Theobroma cacao]	–	–	–	37.3746	49.0971	9.10939	33.0598	1.44551	90.1009	44.4268
Ai02g27325	mitogen-activated protein kinase kinase, partial [Genlisea aurea]	–	–	–	54.7773	64.0532	23.2355	58.5083	26.5917	5.91967	31.1346
Ai02g6749	Uncharacterized protein TCM_018856 [Theobroma cacao]	–	–	–	27.5608	16.2388	9.38645	14.4462	0	133.891	61.6475

(continued)

Table 9.3 (continued)

Gene ID	BLAST description	GO description	KEGG description	EC description	Mature leaf	Flower and bud	Developing endosperm	Fruit coat and pulp	Mature fruit	Seedling root	Seedling shoot
Ai02g16441	F-box/RNI-like superfamily protein isoform 1 [Theobroma cacao]	"biological_process"	–	–	45.4295	33.0276	33.2482	36.6779	15.0754	48.882	47.7373
Ai02g13538	Uncharacterized protein TCM_029293 [Theobroma cacao]	–	–	–	29.7181	55.2772	28.3721	45.7245	54.0822	12.7567	30.8562
Ai02g17284	Membrin, putative [Theobroma cacao]	"protein transport"	"SNARE interactions in vesicular transport"	–	13.7675	42.992	16.9358	47.637	49.1472	60.4056	25.5298
Ai02g11477	unnamed protein product [Vitis vinifera]	–	–	–	31.532	39.2324	86.98	61.7386	2.37809	24.7051	9.74524
Ai02g4595	PREDICTED: putative F-box/FBD/LRR-repeat protein At1g78840-like [Cucumis sativus]	"biological_process"	–	–	4.68907	22.2632	101.38	41.9257	23.6668	36.7385	23.4102
Ai02g15792	ubiquitin-protein ligase, putative [Ricinus communis] gb|EEF29668.1| ubiquitin-protein ligase, putative [Ricinus communis]	"cytoplasm"	–	–	38.2536	29.5229	36.0605	48.6221	29.6825	36.3116	32.2091
Ai02g17911	predicted protein [Populus trichocarpa]gb|EEE86367.1| hypothetical protein POPTR_0004s09110 g [Populus trichocarpa]	–	–	–	57.8878	54.5078	16.4679	25.9868	14.808	4.77417	70.0982
Ai02g21734	Calcium-binding EF-hand family protein, putative [Theobroma cacao]	–	–	–	76.422	29.9173	4.30626	23.567	0.893585	41.774	62.2514
Ai02g17805	Uncharacterized protein isoform 2 [Theobroma cacao]	–	–	–	2.39875	1.17041	119.372	16.4011	0	90.2114	7.41356
Ai02g10431	predicted protein [Populus trichocarpa]	–	–	–	0	235.843	0	0	0	0	0
Ai02g7323	hypothetical protein PRUPE_ppa021024 mg [Prunus persica]	–	–	–	18.1772	32.7615	99.3563	17.9086	3.3573	45.9224	16.4332
Ai02g5964	PREDICTED: UDP-glycosyltransferase 73C2-like [Glycine max]	–	–	–	62.7797	16.122	45.3673	10.0409	4.43017	20.7105	73.7527

(continued)

Table 9.3 (continued)

Gene ID	BLAST description	GO description	KEGG description	EC description	Mature leaf	Flower and bud	Developing endosperm	Fruit coat and pulp	Mature fruit	Seedling root	Seedling shoot
Ai02g24664	predicted protein [Populus trichocarpa]	–	–	–	14.6624	33.9822	5.01016	66.3282	0	72.7564	37.7628
Ai02g39025	unnamed protein product [Vitis vinifera]	–	–	–	0	11.0917	59.3169	41.4477	80.007	28.497	9.36752
Ai02g231	Cytochrome P450, family 71, subfamily A, polypeptide 25, putative [Theobroma cacao]	"membrane"	–	–	11.9479	83.8663	2.54808	4.4346	3.87786	24.5478	97.6838
Ai02g10171	Uncharacterized protein isoform 1 [Theobroma cacao] gb[EOY24061.1] Uncharacterized protein isoform 1 [Theobroma cacao]	–	–	–	1.96481	29.2397	41.7475	59.6391	31.1183	40.0247	24.2896
Ai02g78	conserved hypothetical protein [Ricinus communis]gb[EEF45471.1] conserved hypothetical protein [Ricinus communis]	–	–	–	35.5757	26.8393	30.4299	39.1037	49.268	28.7526	16.4088
Ai02g8848	Uncharacterized protein TCM_019327 [Theobroma cacao]	–	–	–	9.05967	14.7226	60.1723	33.1045	31.4626	42.2452	32.8036

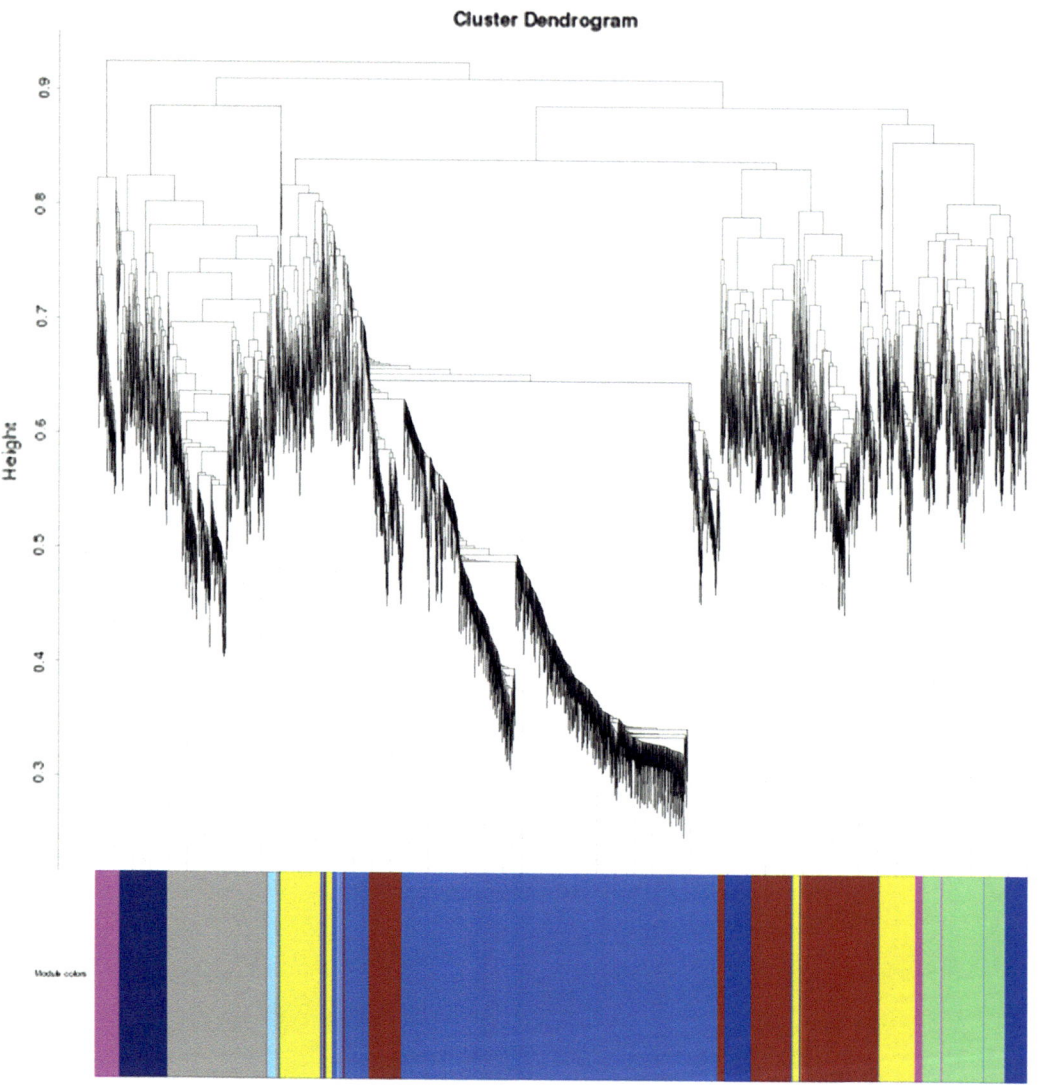

Fig. 9.3 Clustering of genes based on co-expression. The Y-axis represents the height which generalizes the distance between cluster pairs. The different clusters based on co-expression is marked in colors on X-axis

Table 9.4 Grouping of genes based on co-expression

Sl No	Group color	Number of genes
1	Black	2194
2	Blue	3813
3	Brown	2847
4	Cyan	470
5	Dark green	184
6	Dark grey	165
7	Dark orange	112
8	Dark red	184
9	Dark turquoise	174
10	Green	2478
11	Green–yellow	778
12	Grey	14,455
13	Grey60	393
14	Light cyan	431
15	Light green	348
16	Light yellow	281
17	Magenta	1431
18	Midnight blue	432
19	Orange	118
20	Pink	1874
21	Purple	999
22	Red	2311
23	Royal blue	209
24	Salmon	568
25	Tan	762
26	Turquoise	3861
27	White	95
28	Yellow	2528

9.9 Conclusions

The gene prediction using two different programs and selecting common genes yielded complete set of genes which were further supported by gene expression information from RNA-seq data. The annotation process provided annotation for both genomic and organelle genes. The analysis also showed that 3008 genes showing tissue-specific expression in various tissues, while 14,249 genes were expressed in all the tissues.

Table 9.5 Mitochondria genes annotated in *A. indica* mitochondrial genome

Sl No	Gene name	Genes description	Start	End	Strand
1	AiMig01	atp1	59,372	59,941	+
2	AiMig02	atp4	15,040	15,540	+
3	AiMig03	atp6	10,088	9327	−
4	AiMig04	atp8	10,664	10,188	−
5	AiMig05	atp9	14,385	14,164	−
6	AiMig06	ccmB	93,499	92,882	+
7	AiMig07	ccmC	18,311	19,006	+
8	AiMig08	ccmFc	189,803	189,027	−
9	AiMig09	ccmFn	88,211	86,481	−
10	AiMig10	cob	5908	4730	−
11	AiMig11	cox1	22,347	23,072	+
12	AiMig12	cox2	4006	3254	−
13	AiMig13	cox3	17,841	17,047	−
14	AiMig14	matR	211,171	210,569	−
15	AiMig15	mttB	203,616	204,410	+
16	AiMig16	nad1	29,870	30,217	+
17	AiMig17	nad2	36,536	36,384	−
18	AiMig18	nad3	19,948	19,592	−
19	AiMig19	nad4	157,499	157,041	−
20	AiMig20	nad4L	14,560	14,859	+
21	AiMig21	nad5	13,867	13,631	−
22	AiMig22	nad6	6882	6226	−
23	AiMig23	nad7	48,330	48,485	+
24	AiMig24	nad9	144,283	144,846	+
25	AiMig25	rpl2	209,069	208,176	−
26	AiMig26	rpl5	3999	4556	+
27	AiMig27	rpl10	193,222	193,677	+
28	AiMig28	rpl16	169,876	170,430	+
29	AiMig29	rps1	140,247	139,693	+
30	AiMig30	rps2	33,515	33,414	−
31	AiMig31	rps3	168,332	168,409	+
32	AiMig32	rps4	33,449	32,415	−
33	AiMig33	rps7	82,586	83,047	+
34	AiMig34	rps10	105,617	105,402	−
35	AiMig35	rps12	19,539	19,168	−
36	AiMig36	rps14	4561	4620	+
37	AiMig37	rps19	168,202	168,315	+
38	AiMig38	sdh3	97,006	97,299	+
39	AiMig39	sdh4	17,116	16,712	−

Table 9.6 Annotation of chloroplast genes

Sl no	Gene name	Description	Start	End	Strand
1	AiChg01	atpI	478	1035	+
2	AiChg02	rpoC2	1137	2807	+
3	AiChg03	rps2	3970	4599	+
4	AiChg04	rpoC1	30,025	31,467	+
5	AiChg05	rpoB	5793	7298	+
6	AiChg06	lhbA	9592	9777	−
7	AiChg07	atpB	107,779	109,272	−
8	AiChg08	clpP	42,475	42,765	+
9	AiChg09	rpl2	10,814	11,245	+
10	AiChg10	rps19	11,317	11,592	+
11	AiChg11	rpl22	11,641	12,033	+
12	AiChg12	rpl23	67,343	67,576	+
13	AiChg13	ycf2	24,006	26,771	+
14	AiChg14	ndhB	63,848	64,624	−
15	AiChg15	psbK	17,130	17,312	−
16	AiChg16	ycf1	99,366	99,734	−
17	AiChg17	ndhF	20,260	21,384	+
18	AiChg18	rpl32	21,967	22,110	+
19	AiChg19	ccsA	90,390	90,614	+
20	AiChg20	ndhD	91,084	92,124	−
21	AiChg21	ndhG	22,638	22,925	+
22	AiChg22	ndhA	99,567	99,974	−
23	AiChg23	orf188	62,209	62,559	+
24	AiChg24	atpF	27,554	27,817	+
25	AiChg25	atpA	27,961	29,409	+
26	AiChg26	petN	33,763	33,855	−
27	AiChg27	ycf3	34,890	35,018	+
28	AiChg28	accD	36,389	36,961	−
29	AiChg29	cemA	40,947	41,585	−
30	AiChg30	ycf10	38,152	38,295	−
31	AiChg31	psbE	38,423	38,671	+
32	AiChg32	psbF	77,505	77,621	+
33	AiChg33	psbL	38,826	38,939	+
34	AiChg34	psbJ	39,088	39,207	+
35	AiChg35	petA	39,723	40,496	−
36	AiChg36	rps18	41,830	42,051	−
37	AiChg37	petB	43,877	44,326	−
38	AiChg38	psbH	45,160	45,360	−
39	AiChg39	psbN	45,465	45,545	+
40	AiChg40	psbT	45,582	45,680	−

(continued)

Table 9.6 (continued)

Sl no	Gene name	Description	Start	End	Strand
41	AiChg41	psi_psbT	45,806	47,308	−
42	AiChg42	rps12_3end	49,065	49,316	−
43	AiChg43	rps12	49,077	49,316	−
44	AiChg44	ycf15	50,112	50,204	+
45	AiChg45	orf56	55,335	55,421	+
46	AiChg46	orf42	55,699	55,824	+
47	AiChg47	ycf68	56,705	56,884	−
48	AiChg48	rps15	60,727	60,984	+
49	AiChg49	ndhH	61,026	62,204	+
50	AiChg50	rps7	64,943	65,407	−
51	AiChg51	atpE	107,363	107,695	−
52	AiChg52	rpl33	72,528	72,689	−
53	AiChg53	rpl36	73,139	73,249	+
54	AiChg54	rps11	73,370	73,783	+
55	AiChg55	rpoA	74,175	75,275	+
56	AiChg56	petD	75,408	75,950	−
57	AiChg57	rpl20	78,886	79,236	−
58	AiChg58	rpl14	80,215	80,484	−
59	AiChg59	rpl16	80,683	80,904	−
60	AiChg60	rps14	81,388	81,687	−
61	AiChg61	psaB	81,972	84,248	−
62	AiChg62	psaA	85,665	86,681	−
63	AiChg63	petG	88,992	89,069	+
64	AiChg64	rps3	92,869	93,522	−
65	AiChg65	ndhE	93,821	94,009	−
66	AiChg66	ndhJ	94,515	94,976	−
67	AiChg67	ndhK	95,093	95,959	−
68	AiChg68	ndhC	95,842	96,201	−
69	AiChg69	psbA	97,064	98,122	−
70	AiChg70	matK	98,593	98,976	−
71	AiChg71	ycf4	100,559	100,900	+
72	AiChg72	rps4	101,795	102,295	−
73	AiChg73	rbcL	103,617	105,041	+
74	AiChg74	atpH	106,705	106,848	−
75	AiChg75	psbD	109,690	110,706	+
76	AiChg76	psbC	111,084	112,262	+
77	AiChg77	rbcLr	112,787	112,957	−

Acknowledgements We acknowledge Genomics facility (BT/PR3481/INF/22/140/2011) at Centre for Cellular and Molecular Platforms, Bangalore for sequencing of Neem genomes. We acknowledge Pradeep H, Aarati Karaba, Manojkumar S and Annapurna for their help in NGS library preparation and sequencing. We thank Ashmita G and Divya S for their help in manual curation of SSR markers. We are grateful to Rajanna, National Botanical Garden, University of Agricultural Sciences, GKVK campus, Bangalore for his help during neem sample collection.

References

Altschul SF, Gish W, Miller W, Myers EW, Lipman DJ (1990) Basic local alignment search tool. J Mol Biol 215:403–410

Alverson AJ, Wei X, Rice DW, Stern DB, Barry K, Palmer JD (2010) Insights into the evolution of mitochondrial genome size from complete sequences of *Citrullus lanatus* and *Cucurbita pepo* (Cucurbitaceae). Mol Biol Evol 27:1436–1448

Burge C, Karlin S (1997) Prediction of complete gene structures in human genomic DNA. J Mol Biol 268:78–94

Davidson RM, Hansey CN, Gowda M, Childs KL, Lin H, Vaillancourt B, Sekhon RS, de Leon N, Kaeppler SM, Jiang N, Buell CR (2011) Utility of RNA sequencing for analysis of maize reproductive transcriptomes. The Plant Genome J 4(3):191

Delcher AL, Harmon D, Kasif S, White O, Salzberg SL (1999) Improved microbial gene identification with GLIMMER. Nucleic Acids Res 27:4636–4641

Grabherr MG, Haas BJ, Yassour M, Levin JZ, Thompson DA, Amit I, Adiconis X, Fan L, Raychowdhury R, Zeng Q (2011) Full-length transcriptome assembly from RNA-Seq data without a reference genome. Nat Biotechnol 29:644–652

Jiang H, Wong WH (2008) SeqMap: mapping massive amount of oligonucleotides to the genome. Bioinformatics 24:2395–2396

Jiang H, Wong WH (2009) Statistical inferences for isoform expression in RNA-Seq. Bioinformatics 25:1026–1032

Kent WJ (2002) BLAT—the BLAST-like alignment tool. Genome Res 12:656–664

Kuravadi NA, Yenagi V, Rangiah K, Mahesh HB, Rajamani A, Shirke MD, Russiachand H, Loganathan RM, Lingu CS, Siddappa S (2015) Comprehensive analyses of genomes, transcriptomes and metabolites of neem tree. PeerJ 3:e1066

Langfelder P, Horvath S (2008) WGCNA: an R package for weighted correlation network analysis. BMC Bioinformatics 9:559

Li W, Godzik A (2006) Cd-hit: a fast program for clustering and comparing large sets of protein or nucleotide sequences. Bioinformatics 22:1658–1659

Mudge JM, Harrow J (2016) The state of play in higher eukaryote gene annotation. Nat Rev Genet 17:758

Schmid R, Blaxter ML (2008) annot8r: GO, EC and KEGG annotation of EST datasets. BMC Bioinf 9:180

Stanke M, Keller O, Gunduz I, Hayes A, Waack S, Morgenstern B (2006) AUGUSTUS: ab initio prediction of alternative transcripts. Nucleic Acids Res 34: W435–W439

Wyman SK, Jansen RK, Boore JL (2004) Automatic annotation of organellar genomes with DOGMA. Bioinformatics 20:3252–3255

Comparison of Gene Families and Synteny Analysis from Neem Genome

10

Nagesh A. Kuravadi and Malali Gowda

Abstract

Analysis of gene families is helpful in understanding the gene evolution in species. Here we have studied the orthology of neem genes and constructed the phylogenetic tree based on the gene content in neem genome and comparing it with genes from 23 other plant species using proteinortho. The analysis shows 5832 unique genes with expression in various neem tissues. Following gene family analysis, Synteny analysis was performed which provides relative order of genes in the genome and provides information on genome structure. We aligned citrus (*Citrus sinensis*) and Arabidopsis genomes with neem contigs using MUMMER program. This identified 24,902 anchored neem contigs which contribute to 161 Mb (62%) of annotated neem genome covering 48% of the citrus genome.

The synteny analysis showed a good synteny between neem and citrus and lowest synteny between neem and Arabidopsis.

10.1 Introduction

Phylogenetic reconstruction aims at finding plausible hypotheses of the evolutionary history of genes or species based on genomic sequence information (Hellmuth and Wieseke 2016). The process involves aligning genes to each other based on sequence similarity and classifying genes into families. The process is done for genes within neem and also other plant species. The resulting orthology inference of genes is very useful in understanding the evolution of genes and classifying and annotating them in databases (Cosentino and Iwasaki 2018). Here we have studied the orthology of neem genes and constructed the phylogenetic tree based on the gene content in neem genome at whole-genome scale.

Following gene family analysis, Synteny analysis was performed which provides relative order of genes in the genome and provides information on genome structure (Bhutkar et al. 2006). Synteny provides a framework in which conservation of homologous genes and gene order is identified between genomes of different species (Liu et al. 2018). Synteny analysis helps in understanding the evolutionary history of

N. A. Kuravadi
Centre for Cellular and Molecular Platforms,
National Centre for Biological Sciences, Bengaluru,
Karnataka, India
e-mail: alwaysnagesh@gmail.com

M. Gowda (✉)
Center for Functional Genomics and
Bio-Informatics, The University of
TransDisciplinary and Health Sciences,
Bengaluru, India
e-mail: malalig@tdu.edu.in

genes and genomes across broad phylogenetic groups and divergence times (Zhao and Schranz 2017). We have used citrus as the closest species to neem with a complete genome to analyze the synteny and order the contigs of neem into blocks of citrus genome.

10.2 Gene Family Classification and Phylogenetic Analysis

Analysis of gene families is helpful in understanding the gene evolution in each species and also identifies any unique gene families present in the genome. To perform this analysis, a total proteomes of 23 plant species along with neem were used to detect the homologues and unique gene set. All-vs-all BLAST-P (E-value e^{-10}) was done using Proteinortho program (Lechner et al. 2011). Comparative data from BLAST analysis was further classified into list of potential orthologs, co-orthologs, and paralogs. The program was also grouped the proteins into specific groups by clustering the gene-pairs. The gene content at the ancestral nodes along with the branches leading to these nodes was reconstructed by using Wagner Parsimony and Likelihood-based approaches in the program count (Csuos 2010).

10.3 Gene Ortholog Analysis

The dendrogram constructed using the overlap of gene families among the plant species (Fig. 10.1). The Sweet orange (*C. sinensis*) genome was showed close related to neem (Xu et al. 2012). From the taxonomic perspective, citrus belonging to Rutaceae, which is closest to neem, as both these plants belong to the class

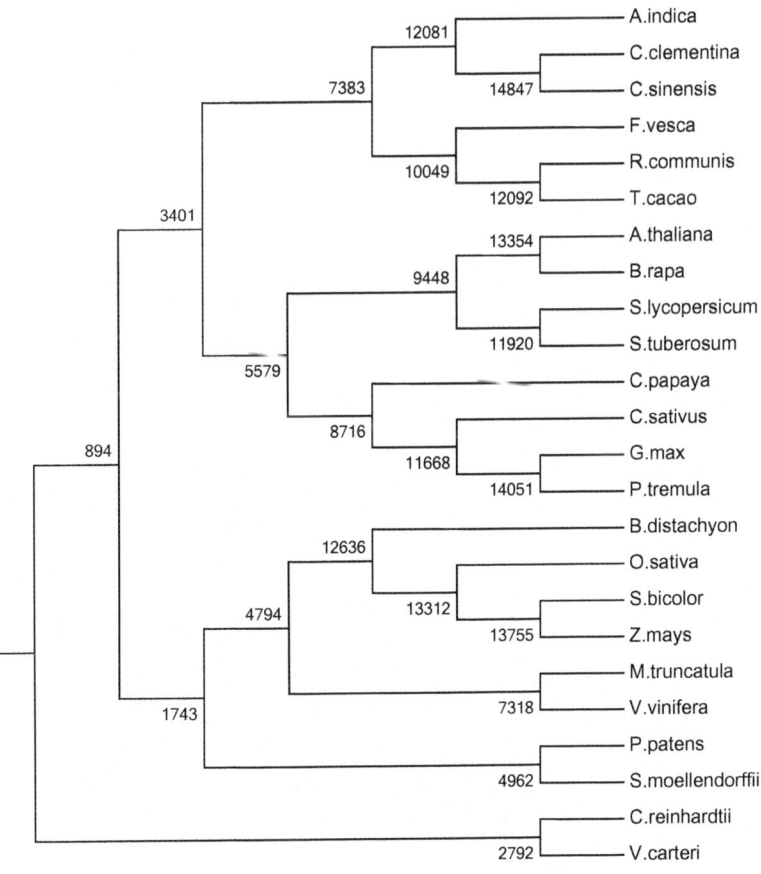

Fig. 10.1 UPGMA species tree based on (co-)orthologous groups detected by Proteinortho. The clustering is based on similarity of the proteome among the plant species

Sapindales. Citrus was the closest member to neem where complete genome information was available for comparative analysis. The orthogroup reconstruction shows that 5122 genes have gained and 2755 genes have lost in comparison with ancestral nodes that based on the gene family expansion in the neem genome (Kuravadi et al. 2015).

Along with genes classification into orthogroups, the BLAST results of protein ortho (Lechner et al. 2011) showed 24,216 genes (54.42%) that were common between neem and citrus genomes. However, 20,279 and 21,931 genes were unique to neem and citrus, respectively. Out of 20,279 unique genes in neem, 5832 genes expressed in various neem tissues (Table 10.1).

Similarly, neem orthologous genes were compared with dicot (Arabidopsis, Populus and grapes) monocot (rice) plant genomes (Fig. 10.2). Nearly 27,498 genes out of 44,495 protein-coding genes from neem were found to have orthology in other plant species with more than 60% identity. Interestingly 38% (16,997) protein-coding genes from neem did not show any orthology with any sequenced plant genomes. However, the number of orthologs in citrus is much higher in comparison to neem possibly due to unique gene families

Table 10.1 Ancestral orthogroup reconstruction using a birth-death model that allows for lineage specific gain/loss rates, as implemented in the program count

Node	Family	Multimember	Gains	Losses	Expansions	Contractions
A. indica	18,327	4320	5122	2755	1128	404
C. clementina	18,998	6643	1967	1344	2625	412
C. sinensis	21,231	9442	5376	2519	3842	644
F. vesca	15,253	3631	3927	3767	872	292
R. communis	16,153	2569	2883	1995	807	211
T. cacao	19,461	9147	5902	1705	4950	118
A. thaliana	17,398	8127	2567	1360	3175	597
B. rapa	17,516	9544	3164	1838	4148	568
S. lycopersicum	17,106	5209	2129	1276	2073	316
S. tuberosum	19,778	10,343	7039	3514	3919	436
C. papaya	14,392	2076	4002	4158	464	330
C. sativus	15,963	5940	3551	2934	2543	437
G. max	23,311	17,969	8293	2245	7225	384
P. tremula	23,987	16,632	8715	1990	6502	575
B. distachyon	17,078	5861	2489	1578	2267	419
O. sativa	19,221	8720	4742	2476	3342	559
S. bicolor	19,054	7922	2904	1517	3056	598
Z. mays	22,592	13,210	8036	3110	5090	651
M. truncatula	16,163	7094	7809	3971	1935	168
V. vinifera	14,536	2915	4040	1829	1097	110
P. patens	11,096	6745	4302	3072	2758	172
S. moellendorffii	9909	3671	3301	3257	993	547
C. reinhardtii	5121	2080	2120	408	548	17
V. carteri	4139	1005	998	268	386	9.5

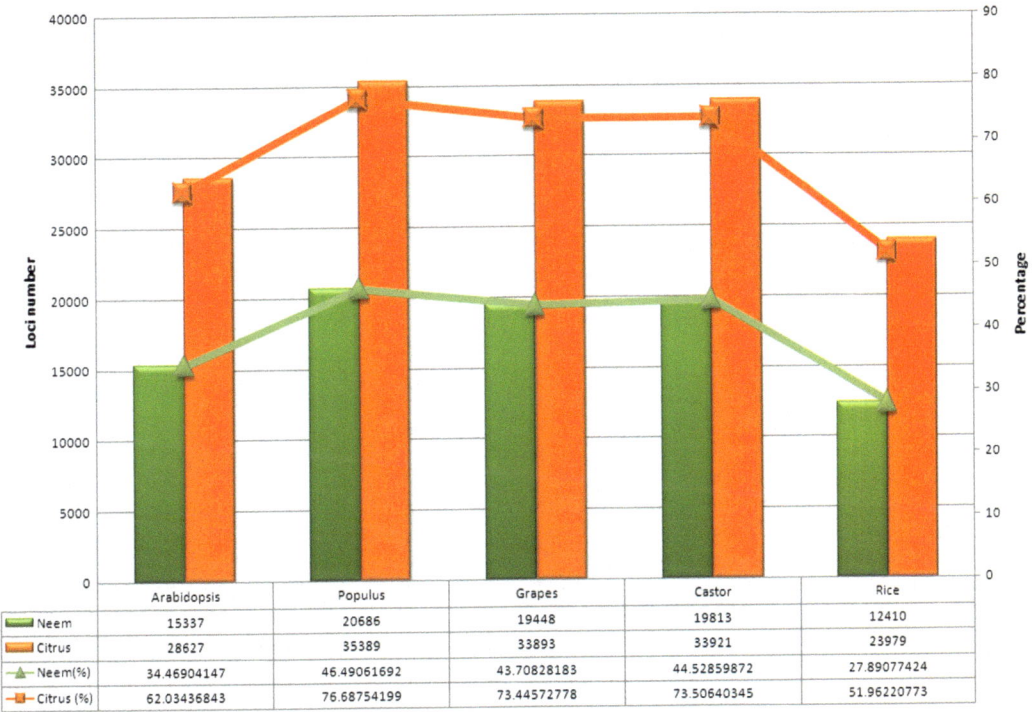

Fig. 10.2 Neem genes with homology with other plant genomes

belonging to Meliaceae/neem. The unique genes in neem (16,997) were further filtered for repeat element presence and expression. This indicates that neem repeat derived genes are active in various tissues. More than 3000 genes found to have no repeat elements and showed expression in various tissues which supports that presence of unique gene families in neem. Among these unique genes, 680 genes share minor homology with hypothetical/predicted proteins in other plant species while 2343 genes found to have no sequence similarity to either predicted or known genes.

10.4 Synteny Analysis

Synteny analysis was done to obtain conserved chromosomal blocks of neem genome and genes in a closely related species. We aligned citrus (*C. sinensis*) and Arabidopsis genomes with neem contigs using MUMMER program (Kurtz et al. 2004). Synteny was computed among neem, citrus and Arabidopsis using Symap4.0 (Soderlund et al. 2011). The synteny information was visualized using the Perl script provided with Symap4.0. This analysis revealed that citrus genome was the closest to neem as expected from phylogeny, so we used citrus genome as a reference to order neem scaffolds. This anchoring analysis method generated nine neem chromosomal segments or pseudo-molecules on citrus chromosomes and one unanchored chromosomal segment. The identified 24,902 anchored neem contigs which contribute to 161 Mb (62%) of annotated neem genome which covering of 48% of the citrus genome. The anchoring had 497 syntenic blocks with 12,176 synteny hits with citrus genome (Kuravadi et al. 2015). Further, synteny analysis was performed by comparing neem, citrus, and Arabidopsis (Fig. 10.3). The synteny analysis showed a good synteny between neem and citrus and lowest synteny between neem and Arabidopsis.

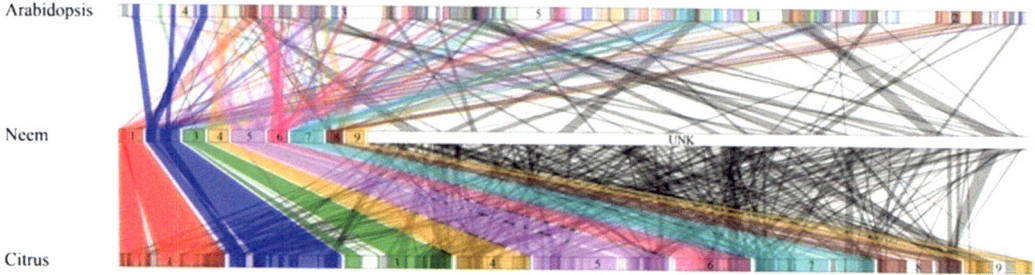

Fig. 10.3 Synteny map of neem genome with citrus and Arabidopsis genomes

Acknowledgements We acknowledge Genomics facility (BT/PR3481/INF/22/140/2011) at Centre for Cellular and Molecular Platforms, Bangalore for sequencing of Neem genomes. We acknowledge Pradeep H, Aarati Karaba, Manojkumar S and Annapurna for their help in NGS library preparation and sequencing. We thank Ashmita G and Divya S for their help in manual curation of SSR markers. We are grateful to Rajanna, National Botanical Garden, University of Agricultural Sciences, GKVK campus, Bangalore for his help during neem sample collection.

References

Bhutkar A, Russo S, Smith TF, Gelbart WM (2006) Techniques for multi-genome synteny analysis to overcome assembly limitations. Genome Informatics 17:152–161

Cosentino S, Iwasaki W (2018) SonicParanoid: fast, accurate and easy orthology inference. Bioinformatics 35:149–151

Csuos M (2010) Count: evolutionary analysis of phylogenetic profiles with parsimony and likelihood. Bioinformatics 26:1910–1912

Hellmuth M, Wieseke N (2016) Construction of gene and species trees from sequence data incl orthologs, paralogs, and xenologs. arXiv preprint. arXiv:160208268

Kuravadi NA, Yenagi V, Rangiah K, Mahesh HB, Rajamani A, Shirke MD, Russiachand H, Loganathan RM, Lingu CS, Siddappa S (2015) Comprehensive analyses of genomes, transcriptomes and metabolites of neem tree. PeerJ 3:E1066

Kurtz S, Phillippy A, Delcher A, Smoot M, Shumway M, Antonescu C, Salzberg S (2004) Versatile and open software for comparing large genomes. Genome Biol 5

Lechner M, Findeiß S, Steiner L, Marz M, Stadler PF, Prohaska SJ (2011) Proteinortho: detection of (Co-)orthologs in large-scale analysis. BMC Bioinf 12:124

Liu D, Hunt M, Tsai IJ (2018) Inferring synteny between genome assemblies: a systematic evaluation. BMC Bioinf 19:26

Soderlund C, Bomhoff M, Nelson WM (2011) SyMAP v3 4: a turnkey synteny system with application to plant genomes. Nucleic Acids Res 39:e68-e68

Xu Q, Chen LL, Ruan X, Chen D, Zhu A, Chen C, Bertrand D, Jiao WB, Hao BH, Lyon MP (2012) The draft genome of sweet orange (*Citrus sinensis*). Nat Genet

Zhao T, Schranz ME (2017) Network approaches for plant phylogenomic synteny analysis. Curr Opin Plant Biol 36:129–134

11 Neem Tissue Culture

Divya Mohan, Ashmita J. Tontanahal,
B. N. Sathyanarayana and Malali Gowda

Abstract

The neem tree is a well-known source of metabolites used in the production of bio-pesticides and medicines. To understand the nature and composition of these metabolites in various tissues, in vitro calli were established. Calli from leaves, petals, cambium, and endosperm of the neem tree were successfully induced. The callusing response from petal and endosperm explants were fast demonstrating callus growth in three days and seven days, respectively. Cambium explant showed slow response taking about thirty days for callus formation. Regenerated plantlets were developed from endosperm explant, a triploid tissue, which contained the highest metabolites among selected explants. The concentration of important metabolites including Azadirachtin, Nimbin, Salanin, Azadiradione, and Epoxy/Hydroxy-azadiradione in the different explants, respective calli and regenerated plants were estimated using Ultrahigh Performance Liquid Chromatography-Mass Spectrometry (UHPLC-MS) method. Interestingly, the leaves of regenerated plant derived from endosperm callus showed over hundred-fold higher Azadirachtin content as compared to leaves of the mature tree. Triploidy of the endosperm derived regenerated plant was confirmed using flow cytometry. Using leaves from these tissue culture-derived triploid plants could be a non-destructive source for commercial extraction of metabolites and producing bio-pesticides.

11.1 Introduction

Neem (*Azadirachta indica*) is considered one of the most versatile evergreen trees and its products have been used for centuries in the field of agriculture and medicine. Neem tree synthesizes secondary metabolites in the form of an array of natural complex compounds called limonoids (Singh et al. 2009). These compounds can synergistically produce a range of effects against pathogens (bacteria and fungi) and pests (Koul et al. 1990). Over 200 limonoid compounds isolated from various parts of the neem tree are known to be effective on a wide range of insect pests. The principal compound Azadirachtin, a tetranortriterpenoid, is a potent insecticide and antifeedant (Arnason et al. 1985, Isman et al. 1990 and Kreutzweiser et al. 2002). It has growth

D. Mohan · M. Gowda (✉)
The University of TransDisciplinary and Health Sciences, Bengaluru, Karnataka, India
e-mail: malalig@tdu.edu.in

D. Mohan · A. J. Tontanahal · B. N. Sathyanarayana
Department of Horticulture, University of Agricultural Sciences, Bengaluru, Karnataka, India
e-mail: sathya.bn@gmail.com

disrupting effect against several pests (Chaturvedi et al. 2004; Singh et al. 2009) and acts on many insect larvae by mimicking the ecdysone hormone that regulates pupation of insects (Mitchell et al, 1997). Other major metabolites occurring in substantial quantities with natural pesticidal properties include Nimbin, Salanin, Epoxy/Hydroxy-azadiradione, and Azadiradione (Sidhu et al. 2004 and Govindachari et al. 1995) (Table 11.1).

These metabolites have low toxicity Broughton et al. (1986); Ley et al. 1993). Hence, for the consistent production of Azadirachtin and other important secondary metabolites, in vitro tissue culture techniques were evaluated (Allan et al. 1994; Srivastava and Chaturvedi 2011). Tissue culture allows the controlled production of secondary metabolites, hence enabling quantification and profiling of the compounds (Hussain et al. 2012). In this chapter the method of generating calli, which are unorganized cell masses, using neem explants: leaf, petal, cambium, and endosperm is presented. The metabolites in explants and respective calli were quantified using Ultrahigh Performance Liquid Chromatography-Mass Spectrometry (UHPLC-MS) (Kuravadi et al. 2015) to evaluate them as a source for extraction and expand the understanding of neem biology. Regeneration experiments were conducted on the callus derived from endosperm, a triploid tissue, owing to the presence of high metabolite concentrations in the explant (Thomas and Chaturvedi 2008). Interestingly, the leaves of the regenerated plant derived from endosperm calli showed over hundred-fold higher Azadirachtin concentration in comparison to the diploid leaves of the mature tree. This regeneration method could, therefore, be used to produce plants for commercial extraction of metabolites throughout the year constructively with low variation.

11.2 Generation of Calli from Neem Explants

Plant tissues can develop into unorganized cell masses know as callus in response to biotic or abiotic stress as well as to wounding. Cells of the callus are totipotent and have the potential to regenerate into an entirely new plant (Ikeuchi et al. 2013). Tissue culture exploits this property of plant cells and tissues, making the production of hundreds and thousands of plants from a small piece of tissue or a single plant material possible. It requires a relatively short time period and space under controlled conditions irrespective of season and weather to produce plants on a year-round basis (Akin-Idowu et al. 2009). Nutrient media with varied standardized concentrations of growth hormones, vitamins, carbon sources, and mineral nutrients are necessary to convert plant tissue into a callus.

11.2.1 Explant Preparation

A 50-year old mature neem tree referred to as "mother plant", located at Botanical Garden, University of Agricultural Sciences, Bengaluru in India was chosen for the collection of plant tissue (Fig. 11.1a). Leaf, cambium, petal, and endosperm were collected from this mother plant for analysis (Fig. 11.1b–e). Leaves, cambium,

Table 11.1 Principal effect of major neem metabolites

Metabolite	Principal effect	References
Azadirachtin	Insecticide	Isman et al. (1990)
Nimbin	Anti-inflammatory	Sidhu et al. (2004)
Salanin	Insect repellent	Satdive et al. (2006)
Azadiradione	Insect feeding inhibitor	Govindachari et al. (1995)
Epoxy/Hydroxy-azadiradione	Insect antifeedant	Govindachari et al. (1996)

Fig. 11.1 The **a** mature neem tree, called "mother plant" used for explant collection. The explants collected were **b** leaf; **c** petal; **d** cambium; and **e** endosperm

and petals were collected during the months of April and May. The endosperms were extracted from seeds of fruits, which were collected early in the month of June. The endosperm, a triploid tissue, was separated from the immature seeds at the primitive dicotyledonous stage of embryo development (Chaturvedi et al. 2004). All explants were subjected to surface sterilization and the plant materials were prepared for inoculation according to the procedure described in Ramamurthy et al. (2012).

11.2.2 Callus Inoculation and Subculture

Full strength solid Murashige and Skoog (MS) media (1962) with different concentrations of growth hormones for each explant were prepared (Table 11.2). The sterilized explants were surface injured aseptically and inoculated in tissue culture bottles containing 30 mL of their respective growth media. Leaf and cambium explants were inoculated on MS media with 2 mg/L 6-Benzylaminopurine (BAP) + 1 mg/L Kinetin (Kn) + 2 mg/L 2,4 Dinitrophenol (2,4-D) and 2 mg/L BAP + 1 mg/L 1-naphthaleneacetic acid (NAA), respectively (Ramamurthy et al. 2012). Petal media was standardized using varying concentrations of Thidiazuron (TDZ) and 2,4-D based on trial and error in triplicates. Full strength MS with 1 mg/L TDZ, 2.5 mg/L 2,4-D, 45 mg/L Ascorbic acid, and 15 mg/L Adenine Sulphate resulted in callus production from petal explant. The endosperm media was derived by taking the media described in Chaturvedi et al. (2004), 0.9 mg/L NAA + 0.35 mg/L BAP and 500 mg/L casein hydrolysate (CH) as the mean value. Nine combinations of low, medium and high concentrations of NAA (0.8, 0.9, 1.0) and BAP (0.25, 0.35, 0.45) with 500 mg/L CH were tested to analyze callus yield. Ten replicates of each explant were inoculated in respective media and maintained in a growth chamber at 25 ± 2 °C with 55 ± 5% of relative humidity for 16 h photoperiod to allow the formation of callus tissue.

Wounding and inoculation into nutrient media under controlled conditions initiated callus generation from the explants. Callus development started with discoloration of the explant, followed by swelling and clumping of the tissue to form a pale mass of cells at injured areas. Callus proliferation occurred in all inoculated bottles for petal, cambium and leaf explants. Petal explants developed callus within 3 days (Fig. 11.2b), which was the fastest callusing response observed in the study. Cambium had the slowest callusing response where discoloration on explant appeared only after 30 days of

Table 11.2 Callus initiation media for explants (Volume = 1 L)

Composition	Leaf	Petal	Cambium	Endosperm
Water (mL)	600	600	600	600
Sugar (g)	30	30	30	30
MS stock solution				
Stock A (mL)	100	100	100	100
Stock B (mL)	10	10	10	10
Stock D (mL)	10	10	10	10
Stock C (mL)	10	10	10	10
Plant growth regulators and gelling agent				
BAP** (mg)	2	–	2	0.45
2,4-D** (mg)	2	2.5	–	–
Kn**	1	–	–	–
NAA** (mg)	–	–	1	0.80
TDZ**	–	1	–	–
Vitamin C (mg)	–	45	–	–
Adenine sulphate (mg)	15	15	30	–
CH** (mg)	–	–	–	500
pH	5.68	5.67	5.68	5.68
Agar (g)	4.6	4.5	4.7	4.8

Make up the volume to 1 L
**BAP- 6-Benzylaminopurine; 2,4 D- 2,4 Dinitrophenol; Kn- Kinetin; IPA—Indole-3-propionic acid; NAA- 1-naphthaleneacetic acid; TDZ—Thidiazuron; CH—Casein Hydrolysate

inoculation (Fig. 11.2c). Cambium tissue also yielded very small callus. The response of leaf explant was visible in 7 days with darkening of the edges and curling at the borders (Fig. 11.2a) to form the mass of cells. Among the nine different trialed media endosperm explants inoculated on MS with 0.8 mg/L NAA (low), 0.45 mg/L (high) BAP and 500 mg/L CH elicited maximum callus productivity (in terms of response time and callus yield) (Fig. 11.2d). Light green masses of calli were formed within 7 days of inoculation in 90% of inoculated bottles. The distinction in natural aging time of the respective plant materials and type of tissues that make up the explant could possibly explain the varying callus responses (Table 11.3).

The differences in response times between the four explants were documented (Table 11.3). Calli from explants were subcultured in fresh media every fourth week. The calli generated from the leaf, cambium and petal explants were maintained and used for metabolite quantification. Endosperm calli (Fig. 11.2d) were used for regenerative studies due to the presence of high metabolite content in the explant (Table 11.4).

11.2.3 Regeneration of Triploid Neem Plant

Regeneration studies on endosperm tissue were carried out with the hypothesis that tissue culture of triploid tissue (endosperm) will generate triploid calli, which will regenerate into triploid plantlets. Endosperm had the highest metabolite content compared to other explants (Table 11.4), indicating the possibility of regenerating plantlets with increased metabolite content. The subcultured endosperm calli (Fig. 11.3a) were transferred on shoot regeneration media (MS media with 1.12 mg/L BAP) (Chaturvedi et al. 2004). Over 90% of calli showed the formation of 1 cm

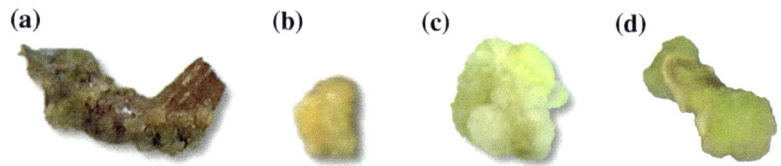

Fig. 11.2 Developed callus formed by inoculation of **a** leaf, **b** petal, **c** cambium, and **d** endosperm explants on respective nutrient media and maintained in a growth chamber at 25 ± 2 °C with 55 ± 5% of relative humidity, for 16 h photoperiod

Table 11.3 Comparing callusing response across neem explants

Explant	No. of days					
	Media	3	7	15	25	45
Leaf	MS + 2 mg/L BAP + 1 mg/L Kn + 2 mg/L 2,4-D	− −	−	+	+	++
Cambium	2 mg/L BAP + 1 mg/L NAA	− −	− −	− −	−	+
Petal	MS + 1 mg/L TDZ + 2.5 mg/L 2,4-D	+	+	++	++	++
Endosperm	MS + 0.8 mg/L NAA + 0.45 mg/L BAP + 500 mg CH	−	+	+	++	++

(++) high yield of callus (+) medium yield of callus (−) low yield of callus (− −) no yield of callus

Table 11.4 Neem metabolites quantification from various explants and respectively derived calluses using LC-MS method

Neem explant	Azadirachtin (ng/mg dry-weight of sample)	Nimbin (ng/mg dry-weight of sample)	Salanin (ng/mg dry-weight of sample)	Azadiradione (ng/mg dry-weight of sample)	Epoxy-azadiradione (ng/mg dry-weight of sample)
Metabolites from Neem explants					
Leaf	BDL	18.5	34.8	23.9	243,200
Petal	BDL	BDL	BDL	BDL	1870.4
Cambium	BDL	BDL	BDL	BDL	420.8
Endosperm	831.6	1367	536	3495	65,096
Metabolites from 30-day old calli					
Leaf callus	BDL	BDL	NF	BDL	BDL
Petal callus	BDL	BDL	BDL	1.4	5.07
Cambium callus	ND	BDL	BDL	NF	1.8
Endosperm callus	2.8	0.04	BDL	NF	0.313

BDL—Below detection limit (<0.01 ng/mg dry-weight of sample); NF—Not found

tall single green shoot on the surface within 3 weeks (Fig. 11.3b) while about 10% of shoots formed were pale without signs of further growth and were hence, discarded. The green shoots were aseptically separated from their calli and inoculated on shoot multiplication media (MS media with 0.5 mg/L BAP). Each inoculated shoot developed callus at the basal region (Fig. 11.3c) from which multiple shoots formed in 7 days. These shoots needed to be elongated before transferring into media for root regeneration. To standardize the elongation media, ½ strength MS with 1 mg/L Gibberellic acid (GA) and varying strengths of Indole-3-butyric acid (IBA) (0.1, 0.5, 1.0, 1.5 and 2.0 mg) were prepared. Samples inoculated on media with GA and 0.1 mg/L IBA, elongated to a height of up to 4 cm (Fig. 11.3d) in 14 days. Media with GA and 0.5 mg/L IBA resulted in adventitious root regeneration from inoculated shoots, which was used as the root regeneration media. Higher auxin (IBA) concentrations (1.0, 1.5, and 2.0 mg/L) in media resulted in the development of callus with hairy roots on the surface at the base of the inoculated shoots.

The elongated shoots (4 cm) were transferred onto root regeneration media; ½ MS media supplemented with 1 mg/L GA and 0.5 mg/L IBA. After two weeks, 2 cm roots formed from each shoot, resulting in the regenerated plantlet. The roots were further elongated to obtain plants suitable for acclimatization. Further root elongation was elicited by transferring the plantlets on ½ MS media without plant growth regulators (Wattanawikkit et al. 2011). The resulting regenerated plantlets with about 4 cm shoot and 6 cm developed roots (Fig. 11.3e) were moved to the nursery for acclimatization. In vitro tissue cultures under controlled conditions are exposed to more than 95% relative humidity inside the culture bottles, which generally results in underdeveloped waxy cuticles, and dysfunctional

Fig. 11.3 Regeneration of endosperm-callus derived plantlets after treating **a** callus in **b** shoot regeneration, **c** shoot multiplication, **d** shoot elongation and **e** root regeneration media in the growth chamber at 25 ± 2 °C with 55 ± 5% of relative humidity, for 16 h photoperiod

Fig. 11.4 The regenerated plantlet was transferred from culture bottles to **a** pro trays (primary hardening), maintained at ∼40 °C with high humidity (60 ± 5%), further **b** transferred to mud pots (secondary hardening) where they were exposed to natural surrounding and finally post acclimatization **c** planted in soil

or absent stomata (Gilly et al. 1997). Therefore, in order to improve the development of the cuticle and stomata, the plantlets need to be aided slowly before being transferred to the complex field growth environment. They were first transferred on pro trays with coco peat (Fig. 11.4a) and drenched with fungicide (2% Bavestine). The trays were kept in tunnel growth chambers maintained at ∼40 °C with high humidity and watered every 14 days (primary hardening). Further, the plantlets were transferred into mud pots and exposed to the natural surrounding (secondary hardening) (Fig. 11.4b). The plants were planted in natural soil and maintained over time for further analyses (Fig. 11.4c).

11.3 Genome Size Estimation of Regenerated Plant

Chaturvedi et al. (2004) confirmed the triploid nature of neem plants that were derived from endosperm-callus, hence, establishing the hypothesis of the developed plantlets being triploid. The endosperm-derived neem plantlets presented profuse branching and rooting, which were described by Morinaga and Fukushima (1935) as characteristic for a triploid plant compared to their diploid counterparts further supporting the hypothesis. Flow cytometry was used to estimate and compare genome sizes of the samples to confirm triploid nature of regenerated plant. Young leaves from the regenerated plant and mature tree were collected for genome size estimation. The leaves were chopped using sterilized blade and treated with nuclear isolation buffer (NIB). The samples were stained with 50 µg/mL propidium iodide (PI), treated with RNase and filtered through 30 µm nylon mesh. The samples were processed for ploidy level estimation as per the protocol followed by Hittalmani et al. 2017 and analyzed using Beckman-Coulter FACS at Bengaluru Genomics Center Pvt Ltd.

Pisum sativum was used as an internal reference standard for genome size estimation due to its ease of preparation and stability (Johnston et al. 1999), which had an average DNA content (2C) value of 9.09 and genome size of 4300 Mb (Fig. 11.5a). The genome size was calculated by multiplying 2C value (pg) with 978 Mb (1 pg equivalent value) (Dolezel and Bartoš 2005). The DNA content of diploid neem sample collected was found to be 0.978 g, resulting in a genome size of 478.2 Mb (Fig. 11.5b). The DNA content and genome size of regenerated tissue was 1.59 pg and 777.51 Mb, respectively (Fig. 11.5c), which is nearly three times the genome size (261.458 Mb) (Gowda et al. 2015) recorded for haploid neem tissue. Thus, the regenerated plants were confirmed as triploid and further sampled for metabolite quantification.

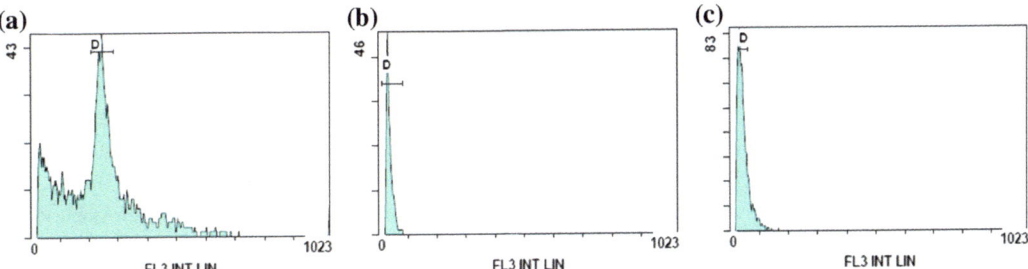

Fig. 11.5 Flow cytometry analysis of **a** internal standard (Pisum sativum), **b** mature neem tree and **c** regenerated plant derived from endosperm callus using Beckman-Coulter FACS. Frequency histograms of the number of nuclei per channel as a function of relative fluorescence in internal standard was plotted. The "*x*" and "*y*" axes represent the number of nuclei and linear fluorescence, respectively

11.4 Quantification of Major Metabolites in Explants and Regenerated Plantlet

Neem extract has been used for its medicinal properties and more extensively in the field of agriculture to produce natural fertilizers for generations. It is environmentally benign, cheap and safe, therefore, has the potential to be used as an alternative to chemical fertilizers. Well-characterized neem metabolites (Azadiractin, Nimbin, Salanin, Azadiradione, and Epoxy/Hydroxy-azadiradione) were quantified in order to understand the distribution of the major metabolites in the plant, the effect of tissue culture on metabolite production and to identify a suitable alternative source for extraction of the major metabolites from neem. Some previous studies on metabolite quantification include analyzing Azadirachtin content in mature leaf (<0.01 ng/mg dry-weight sample) (Kota et al. 2006), 120-day old petal callus (7.6 ng/mg dry-weight sample) (Babu et al. 2006), flower (<0.025 ng/mg dry-weight sample) (Veeresham et al. 1998), bark (<0.001 ng/mg dry-weight sample g) (Wewetzer 1998) and shoots from embryo (8 ng/mg dry-weight sample) (Srividya et al. 1998).

Samples in this study included the explants collected from the mother plant, respective calli and leaves from triploid regenerated plant derived from endosperm-calli. The concentrations of the five metabolites were compared between the leaves of regenerated plant at different stages of development and leaves from the mother plant. The samples were dried at 37 °C for 48 h and powdered with a mortar and pestle using Liquid Nitrogen. The dry powders were stored at 4 °C till metabolite quantification was performed. The UHPLC-MS method established in Kuravadi et al. (2015) was used for quantification of the samples post-treatment, which is a highly sensitive method where lowest concentrations in pico-gram can be detected on the column. The neem metabolites were extracted from the dried powder (2 mg) using 1 mL of methanol followed by 5 min sonication and then centrifuged for 5 min (13,000 rpm, 10 °C); the supernatant was transferred into fresh micro-centrifuge tubes. About 5 μL of supernatant was spiked into 35 μL of methanol along with an internal standard (10 μL Estrone-d4 from 100 μg/mL). For the callus, the methanol extract was completely dried under speed vacuum and reconstituted in 40 μL methanol along with an internal standard (10 μL Estrone-d4 from 100 μg/mL). The analyses were done by injecting 10 μL of the prepared sample into the UHPLC-MS/SRM system (LC-Agilent 1290 Infinity Series, MS—Thermo Fishers TSQ Vantage). The column was used to construct a standard curve in case of Azadirachtin (15.6 pg to 1 ng), Nimbin (3.4 pg to 0.25 ng), Salanin (7.8 pg to 0.5 ng), and Azadiradione (31.25 pg to 2 ng). The final quantification was calculated based on the constructed standard curve (ratio vs. concentration) of individual metabolites.

The concentration of the metabolites was highest in endosperm (831 ng/mg dry weight of sample), while present below detectable limits (BDL) (<0.01 ng/mg dry weight of sample) in the other explants (Table 11.4). Epoxy/Hydroxy-azadiradione was the most abundant metabolite in the endosperm explant, followed by Azadirachtin, Salanin, Azadiradione, and Nimbin (Table 11.4). The high metabolite content in the endosperm could be present as natural protection from insect pests. The metabolites were also BDL in generated calli of all explants, excluding the use of these tissues as sources for extraction. Even after subsequent subcultures, there was no increase in metabolite quantities (Fig. 11.6). After the third subculture (120 days), metabolites were detectable only in the endosperm calli, however, in a low amount.

Regeneration studies were carried out on endosperm calli with the hypothesis of producing plantlets having a high concentration of metabolites compared to the other explants. The leaves from regenerated plants after 60 days (growth room), 120 days (post-primary hardening), 240 days (post-secondary hardening) and 365 days (post-acclimatization) of development (Fig. 11.3a) were sampled for metabolite quantification. There was over ten-fold increase in Azadirachtin content in regenerated leaf (11.42 ng/mg dry-weight of sample) from the 60 days old plantlet in comparison to the mature leaf from the mother plant (<0.01 ng/mg dry-weight of sample). After 120 days (post-primary hardening) hundred-fold increase in Azadirachtin content (248 ng/mg dry-weight of sample) was found with an additional two-fold increase (477 ng/mg dry-weight of sample) after 240 days (post-secondary hardening). Nimbin and Salanin were present in lower quantities in the regenerated leaves compared to the explant. Epoxy/Hydroxy-azadiradione, which is known to have an inferior antifeedant effect compared to Azadirachtin (Govindachari et al. 1996), was present in large quantities in the regenerated plants. Its concentration in the regenerated leaves (714 µg/mg dry-weight of sample) exceeded the naturally occurring seeds (endosperm) by ten-folds (Fig. 11.7). In such high concentrations, it could potentially be a viable metabolite for commercial extraction from the leaves of the regenerated plants. Azadiradione, which was present in high amounts in the leaves of the mature mother plant (23.87 ng/mg dry-weight of sample), was absent (not found, NF) in 60-day-old regenerated leaves grown under greenhouse conditions. However, the leaves from the older regenerated plant showed its presence (from NF to BDL). Conversely, Azadirachtin, Nimbin, and Salanin initially showed an increased presence in regenerated plantlet and were then found to decrease in quantity as the regenerated plant aged. There could be a possible link between the aging of the plant, decline of Azadirachtin and increase in Azadiradione. The possible explanation could be the need plants at an early stage have for the more potent Azadirachtin as a chemical protection against insects-pests compared to a mature tree (Aerts

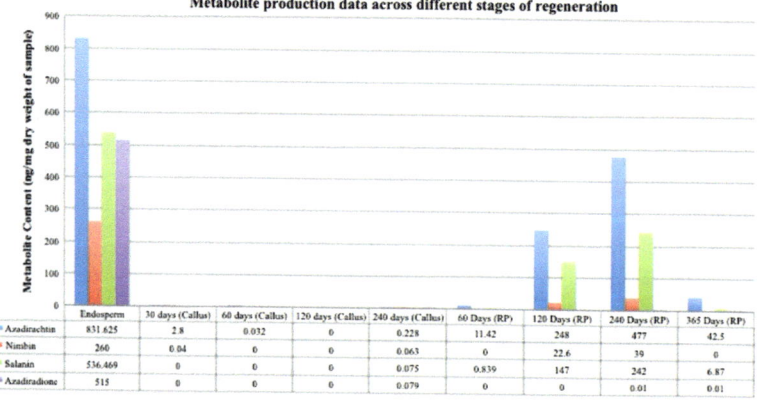

Fig. 11.6 Quantification of major metabolites using the UHPLC-MS explant, calli and regenerated plant (RP)

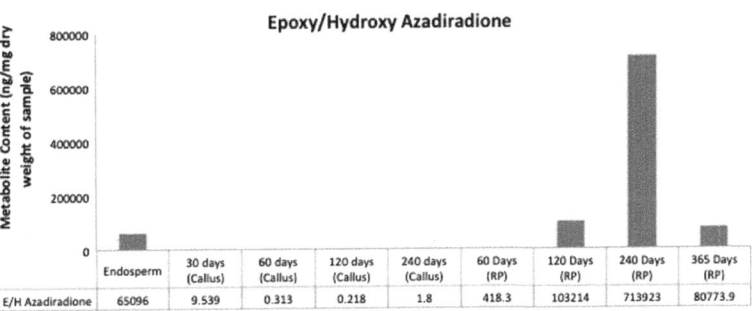

Fig. 11.7 Presence of epoxy/hydroxy Azadiradione across endosperm, explant, calli and regenerated plant (RP)

and Mordue (Luntz) 1997). As hypothesized, the plantlets from the triploid tissue having naturally high metabolite concentrations also showed a high presence of the metabolites that could be used for extraction.

11.5 Endosperm Derived Plantlets—Future Source for Metabolite Extraction

With low variation and higher availability, leaves from triploid regenerated plants derived from endosperm-callus could be a reliable source for isolation of phytochemicals and secondary metabolites. The amount of metabolites in the endosperm (seeds) is higher than in the leaves of regenerated plantlets, however, the endosperm cannot be used extensively for metabolite extraction. Therefore, as an alternative, regenerated leaves can be harvested for the production of bio-pesticides for farming throughout the year. Further, tissue culture excludes the geographical restriction to a large extent and allows more accessible research to be carried out. This study provided important metabolite data to decode the biosynthetic pathways of different neem metabolites, know the interaction and dependence among each other. For example, the influence of Azadirachtin, Nimbin, and Salanin on the production of Azadiradione during the different stages of plant development could be studied based on the variation in concentration. Further research on this area could provide data on the functionality of the different metabolites and methods to maximize their production. The technique could be extended to other important medicinal plants.

11.6 Conclusion

This study is the first of its kind to compare levels of production of secondary metabolites between explant material and respective calli from leaf, cambium, petal, and endosperm. The regenerated triploid plants derived from endosperm calli were successfully established in the soil, allowing further studies on the neem metabolome. The leaves of these plants contain relatively high metabolite quantities including Azadirachtin (>400 ng/mg dry-weight of sample), which could be used for commercially extraction. Tissue-cultured plants can be mass multiplied for large-scale plantation. This could result in continuous production and supply of neem metabolites, thus giving rise to a non-destructive approach to explore neem biodiversity without geographical limitations.

References

Aerts RJ, Mordue (Luntz) AJ (1997) Feeding deterrence and toxicity of neem triterpenoids. J Chem Ecol 23(9): 2117–2132. Available at: http://link.springer.com/10.1023/B:JOEC.0000006433.14030.04. Accessed 8 Oct 2017

Akin-Idowu P, Ibitoye DO, Ademoyegun OT (2009) Tissue culture as a plant production system for horticultural crops. Afr J Biotechnol 8(16): 3782–3799

Allan EJ, Eeswara JP, Johnson S, Mordue AJ, Morgan ED, Stuchbury T (1994) The production of azadirachtin by in-vitro tissue cultures of neem. Pestic Sci 42:147–152

Arnason JT, Philogene BJR, Donskov N, Hudon M, McDougall C, Fortier G, Morand P, Gardner D, Lambert J, Morris C, Nozzolillo C (1985) Antifeedant and insecticidal properties of azadirachtin to the European corn borer, Ostrinia nubilalis. Entomologia experimentalis et applicata 38(1):29–34

Babu VS, Narasimhan S, Nair GM (2006) Bioproduction of azadirachtin-A, nimbin and salannin in callus and cell suspension cultures of neem (*Azadirachta indica* A. Juss.). Sci Correspondence 91:22–24

Broughton H, Ley SV, Slawin AMZ, Williams DJ, Morgan ED (1985) X-ray crystallographic structure determination of detigloyldihydro azadirachtin and reassignment of the structure of the limonoid insect antifeedant. J Chem Soc Chem Commun, 46–47

Chaturvedi R, Razdan MK, Bhojwani SS (2004) In vitro clonal propagation of an adult tree of neem (*Azadirachta indica* A. Juss.) by forced axillary branching. Plant Sci 166:501–506

Dolezel J, Bartoš J (2005) Plant DNA flow cytometry and estimation of nuclear genome size. Ann Bot 95(1):99–110

Gilly C, Rohr R, Chamel A (1997) Ultrastructure and radiolabelling of leaf cuticles from ivy (Hedera helix Hedera helix L.) plants in vitro and during ex vitro acclimatization. Ann. L. plants in vitro and during ex vitro acclimatization. Ann Bot 80: 139–145

Govindachari TR, Narasimhan NS, Suresh G, Partho PD, Gopalakrishnan G (1996) Insect antifeedant and growth-regulating activities of salannin and other C-seco limonoids from neem oil in relation to azadirachtin. J Chem Ecol 22(8):1453–1461

Govindachari TR, Narasimhan NS, Suresh G, Partho PD, Gopalakrishnan G, Krishnakumari GN (1995) Structure-related insect antifeedant and growth regulating activities of some limonoids. J Chem Ecol 21:1585–1600

Gowda M et al. (2015) Azadirachta indica cultivar GKVK, whole genome shotgun sequencing project. Nat Centre Biotechnol Inf

Hittalmani S et al (2017) Genome and Transcriptome sequence of Finger millet (Eleusine coracana (L.) Gaertn.) provides insights into drought tolerance and nutraceutical properties. BMC Genom 18:1–16

Hussain M, Rahman M, Fareed S, Ansari S, Ahmad I, Saeed M (2012) Current approaches toward production of secondary plant metabolites. J Pharm Bioallied Sci 4(1):10

Ikeuchi M, Sugimoto K, Iwase A (2013) Plant callus: mechanism of induction and repression. Plant cell 25 (9):3159–3173

Isman MB, Koul O, Luczynski A, Kaminski J (1990) Insecticidal and antifeedant bioactivities of neem oils and their relationship to azadirachtin content. J Agric Food Chem 38(6):1406–1411

Johnston SJ et al (1999) Reference standards for determination DNA content of plNT NUCLEI 1. Am J Bot 86(5):609–613

Kota S, Rao RND, Parvathi C (2006) In vitro response of select regions of *Azadirachta indica* A. Juss (Meliaceae) as elucidated by biochemical and molecular variations. Curr Sci 91:770–776

Koul O, Isman MB, Ketkar CM (1990) Properties and uses of neem. *Azadirachita indica*. Can J Bot 68:1–11

Kreutzweiser DP, Back RC, Sutton TM, Thompson DG, Scarr TA (2002) Community-level disruptions among zooplankton of pond mesocosms treated with a neem (azadirachtin) insecticide. Aquat Toxicol 56(4):257–273

Kuravadi NA, Yenagi V, Rangiah K, Mahesh H, Rajamani A, Shirke MD, Russiachand H, Loganathan RM, Shankara LC, Siddappa S, Ramamurthy A, Sathyanarayana B, Gowda M (2015) Comprehensive analyses of genomes, transcriptomes and metabolites of neem tree. PeerJ 3:e1066

Ley SV, Denholm AA, Wood A (1993) The chemistry of azadirachtin. Nat Pro Rep 10:109–157

Mitchell MJ, Smith SL, Johnson S, Morgan ED (1997) Effects of the neem tree compounds azadirachtin, salannin, nimbin, and 6-desacetylnimbin on ecdysonemono oxygenase activity. Arch Insect Biochem Physiol 35:199–209

Morinaga T, Fukushima E (1935) Cytological studies on *Oryza sativa* L. II. Spontaneous autotriploid mutants in *Oryza sativa* L. Jpn J Bot 7:207–225

Murashige T, Skoog F (1962) A revised medium for rapid growth and bio-assays with tobacco tissue cultures. Physiol Plant 3:473–497

Ramamurthy A, Kag B, Hegde V, Loganathan MR, Saiyed T, Sathyanarayana BN, Gowda M (2012) Studies on in vitro regeneration from various explants of a mature neem (*Azadirachta indica*) tree. Acta Hort 961:449–456

Satdive RK, Fulzele DP, Eapen S (2006) Enhanced production of azadirachtin by hairy root cultures of *Azadirachta indica* A. Juss by elicitation and media optimization. J Biotechnol 128:281–289

Sidhu OP, Kumar V, Behl HM (2004) Variability in triterpenoids (nimbin and salanin) composition of neem among different provenances of India. Ind Crops Prod 19(1):69–75

Singh KK, Phogat S, Dhillon RS, Tomar A (2009) Neem. A Treatise, 4–6

Srivastava P, Chaturvedi R (2011) Increased production of azadirachtin from an improved method of androgenic cultures of a medicinal tree *Azadirachta indica* A. Juss. Plant Signaling Behav 6(7):974–981

Srividya N, Sridevi BP, Satyanarayana P (1998) Azadirachtin and nimbin content in in vitro cultured shoots and roots of *Azadiracta indica* A. Juss. Ind J Plant Physiol 3:129–134

Thomas TD, Chaturvedi R (2008) Endosperm culture: a novel method for triploid plant production. Plant Cell Tissue Organ Cult 93(1):1–14

Veeresham C, Kumar MR, Sowjanya D, Kokate CK, Apte SS (1998) Production of azadirachtin from callus cultures of *Azadirachta indica*. Fitoterapia 69: 423–424

Wattanawikkit P, Bunn E, Chayanarit K, Tantiwiwat S (2011) Effect of cytokinins (BAP and TDZ) and auxin (2,4-D) on growth and development of paphiopedilum callosum. Nat Sci 45:12–17

Wewetzer A (1998) Callus culture of Azadirachta indica and their potential for the production of azadirachtin. Phytoparasitica 26:47–52

Neem Microbiome

12

Varalaxmi B. Agasimundin, Kannan Rangiah,
Ambardar Sheetal and Malali Gowda

Abstract

Endophytes are the microorganisms that persist within or between tissues of plants without causing any negative effect on plant's growth and development. Endophytes are mutually co-evolved with host plant species and have been reported to produce secondary metabolites similar to host plants. In this study, we developed unique neem-based media to isolate neem endophytic fungi and bacteria. A total of 361 fungi and 80 bacterial endophytes were isolated from various parts of neem plant including leaf, flower, seed, bark, cortex, and root using neem-based media. This is the first of its kind to demonstrate culturing of endophytic fungi and bacteria on a selective neem-based media without using any external nutrients. Out of total 376 fungal endophytes and 80 bacterial endophytes, only 10 fungi and 3 bacteria were inhibiting the growth of pathogenic rice blast fungi, *Magnaporthe oryzae* which were further identified using ITS and 16S rRNA sequencing. Finally, two fungal isolates identified as *Fusarium sp.,st.*Ai.67A and *Neocosmospora ramose* Ai.51D and two bacterial isolates identified as *Pantoea sp.* Ai.A2 and *Bacillus sp.* Ai.C5 were further selected for secondary metabolite production. These microbes were found to produce three neem metabolites such as Epoxy/hydroxyazadiradione, nimbin, and salanin. These endophytes will have great potential applications in agriculture, medicine, and bio-energy.

12.1 Introduction

12.1.1 Neem Tree

The divine tree neem (*Azadirachta indica* A. Juss) is a tropical evergreen plant native to east India and Burma. It grows abundantly in southeastern Asia and West Africa (Verma and Kharwar 2006). Neem is an omnipotent tree and a sacred gift of nature. Neem tree is mainly cultivated in the Indian subcontinent. The neem tree is a divine tree known for its medicinal and insecticidal properties. It has important applications in pharmaceutical field. The medicinal utilities have been described especially for neem leaf. Neem leaf and its constituents have been demonstrated to exhibit immunemodulatory, anti-inflammatory, anti-hyperglycaemic, antiulcer, antimalarial, antifungal, antibacterial, antiviral, antioxidant, antimutagenic and anticarcinogenic properties (Brahmachari

---

V. B. Agasimundin · A. Sheetal · M. Gowda (✉)
Centre for Functional Genomics and Bioinformatics,
TransDisciplinary University, Bengaluru, India
e-mail: malalig@tdu.edu.in

V. B. Agasimundin · K. Rangiah · A. Sheetal ·
M. Gowda
Centre for Cellular and Molecular Platforms,
NCBS-TIFR, GKVK Post, Bellary Rd,
Bengaluru, India

2004). Medicinal and agricultural applications of neem products have been dated back to Indian civilization.

12.1.2 Plant Endophytes

Endophytes are microbes that colonize the living internal tissues of plants without causing any immediate negative effects on host plant. They are a largely unexplored component of biodiversity, especially in the tropics. Endophytic fungi have been isolated from leaves, stems, and roots of woody plants in the temperate regions and the tropics (Verma et al. 2011). Neem tree is known to harbor several endophytic bacteria and fungi (Rajagopal and Suryanarayanan 2000; Mahesh et al. 2005; Verma et al. 2007). They have a protective role against insect herbivore and many are potential producers of novel antimicrobial secondary metabolites. Endophytes are constantly exposed to intergeneric genetic exchange with the host plant. They are known to produce wide variety of beneficial bioactive compounds similar to their host plant. Isolation of a potent anticancer agent, taxol from *Pestalotiopsis microspora*, endophytes of the yew tree and the phytohormone producing fungus from rice plant, *Gibberella fujikuroi* suggests the potential of endophytes as a source of useful metabolites. Hence, screening of these endophytes has become essential step for the discovery of novel metabolites or bioactive compounds. This chapter focuses on the isolation and characterization of fungal endophytes from the neem plant using various microbiological, genomics and metabolomics techniques.

12.2 Neem Endophyte Isolation, Preservation, and Revival

12.2.1 Isolation of Fungal Endophytes from Neem Explants

A 50 years old neem tree was selected for endophytes isolation from botanical garden located at University of Agricultural Sciences, Bangalore (13.0777° N, 77.5805° E) (Fig. 12.1).

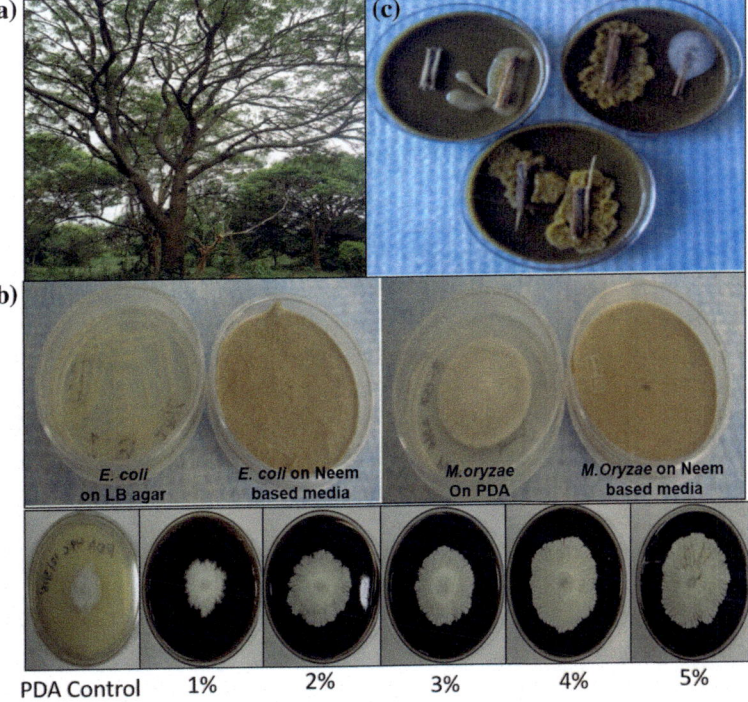

Fig. 12.1 a Endophytic bacteria and fungi isolated from 50 years old neem tree, b growth assay of bacteria (*E. coli*) and rice blast fungus (*M. oryzae*) growth assay on synthetic media and neem-based media, c Isolation of endophytes on neem-based media and d Effect of different concentrations of neem media was checked for the growth of endophytes

Table 12.1 Neem microbes isolated from neem explants

Sl. no	Neem plant parts	No. of plates	No. of bacteria	No. of fungi
1	Outer bark	20	8	93
2	Bark	20	9	23
3	Petiole	20	0	16
4	Cortex	20	17	20
5	Open flower	20	1	23
6	Unopened flower	20	2	10
7	Endocarp (seed coat)	20	7	101
8	Seed bean	20	2	10
9	Leaf	20	24	53
10	Root	20	10	27
11	Total	200	80	376

Table 12.2 Neem media composition

Composition	Reagents (g/L)
Neem-based media	170 g
Agar powder	50 g
Distilled water	1000 ml

The different parts of the neem tree-like leaves, bark, flowers (opened and unopened buds), root and seeds were collected in zip-lock covers and used immediately for endophytes isolation (Table 12.1). A specialized neem media was formulated by washing and drying the neem leaves for a week in sterile condition and crushed into fine powder using a grinder. The media was prepared as per the details mentioned in Table 12.2.

Endophytes were isolated from different parts of the neem tree. In order to avoid the epiphytes, the samples were surface sterilized using the protocol of Verma and coworkers (2011) with some modifications. The tissue was washed with distilled water followed by 70% ethanol, autoclaved distilled water and sodium hypochlorite (100 g per liter) for 2 min each. The final wash was given by autoclaved distilled water to remove excess chemicals and dried in the laminar airflow. Surface sterilized neem explants were cut into small pieces (10–100 mm). About 10 pieces of each explant were inoculated on specialized neem-based media and incubated at room temperature for 5 days for isolation of fungal and bacterial endophyte that are specific to neem tree. Around 376 fungal endophytes and 80 bacterial endophytes were randomly isolated. All the colonies were isolated irrespective of its morphology. Isolated endophytes were purified by streaking five times or more on fresh neem-based medium to obtain pure colonies.

In the present study, the 376 fungal and 80 bacterial endophytes were isolated from 50-years old neem tree using novel neem plant-based medium. All the previous studies have reported isolation of microbes from plant species including neem tree using synthetic media such as PDA, nutrient agar, mycological agar, malt yeast extract agar for fungi and nutrient agar for bacteria (Verma et al. 2007; Strobel 2003; Debbab et al. 2009). PDA medium preferentially allow growth of fungal species such as *Acremoniumacutatum*, *Cladosporium-cladosporioides*, *Curvularialunata,* and Trichoderma sp (Rajagopal and Suryanarayanan 2000; Mahesh et al. 2005; Verma et al. 2007). *Alternariaalternata*, *A. longipes*, and *Aspergillus niger* were isolated more frequently on mycological agar media than other media (Verma et al.

2011). Endophytic fungi, *Fusarium* grew better in apoplastic washing fluid of bean than in the other synthetic media (Schulz et al. 2002). This clearly shows that many synthetic media are non-selective and allow non-specific growth of microbial species. We developed natural neem plant-based medium to isolate bacterial and fungal endophytes. These microbes grow on natural neem medium that are specific to neem tree. There were no external nutrients added this medium.

Over 90% of true endophytes cannot be cultured on a synthetic media, hence a specialized media is required to culture for each plant-associated microbe (Eevers et al. 2015). Therefore, we developed neem plant-based medium using neem tissues, which allowed selective growth of neem endophytes but not any other exogenous microorganisms. Thus, neem plant-based medium served as a selective media (neem ingredients) for isolation of endophytes from neem tissues. Endophytes were selectively isolated from different parts of the neem tree including leaves, petiole, root, bark, flower, cortex, pulp and seed on neem-based medium, which suppress environmental microorganisms (Fig. 12.1b). Other previous studies have isolated endophytic fungi from neem leaves, stem, bark, and fruits (Rajagopal and Suryanarayanan 2000; Mahesh et al. 2005; Verma et al. 2007). However in present study, more fungal endophytes were isolated from seed and bacterial endophytes from leaf (Table 12.1). Similar strategy can be used to isolate microbes from any other plant species by preparing medium from selected plant parts.

12.2.2 Storage and Maintenance of Endophytes

Two methods were followed for storage of the endophytes, glycerol stock method for both fungi and bacteria and dry storage method for fungal isolates. Maintaining and preserving microbes are essential elements of systematic and biodiversity studies. Since, microbes are such a diverse group, several methods of cultivation and preservation are required to ensure their viability along with morphological, physiological, and genetic integrity over time.

A. Glycerol stock

In Glycerol stock method, about 500 µl of sterilized 15% glycerol and 500 µl of sterilized 10% neem media were taken in 2 ml Cryovial tube and mixed properly. 200 µl of endophytic culture was added, mixed and labeled clearly and then stored at −80 °C.

B. Dry storage method

We developed a novel in-vitro long term dry storage method for fungal endophytes using neem leaves that saves significant amount of money and time required for storage. In this

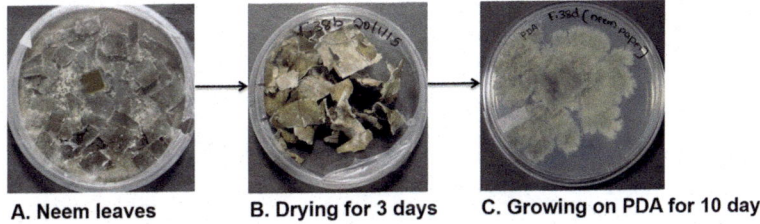

A. Neem leaves   B. Drying for 3 days   C. Growing on PDA for 10 days

Fig. 12.2 Novel storage method of neem endophytes on sterile neem leaves: Dry storage of neem fungal endophytes on neem leaves; **a.** Fungal cultures were inoculated on neem-based media followed by placing the sterile neem leaf pieces in the media plate and allowed to grow at 28 °C for 7 days for the fungus to grow on the leaves. **b.** The inoculated leaf pieces were further air-dried in sterile condition for 3 days at 37 °C and were stored in cryovial at −80 °C. **c.** The culture recovery from −80 °C after 3 years and grown on Potato dextrose agar by incubated at 28 °C for 7 days

method, the fungal endophytes were allowed to grow on Potato Dextrose Agar (PDA) for two days and then autoclaved dry neem leaves were added on fungal endophytes allowing the endophytic fungus to colonize on leaves for 10 days at 28 °C. Petri plates were wrapped with Parafilm to avoid contamination. After full growth of the endophytes on the plates, the neem leaves were taken out and kept for drying in sterile petri plate at 37 °C for 3 days and then the dried leaves were stored in 1.8 ml of Cryovials and stored in −80 °C (Fig. 12.2).

12.2.3 Reviving of Fungal and Bacterial Endophytes

The stored (−80 °C in glycerol) endophytic cultures were collected and revived on neem-based media and then allowed the fungal endophytes to grow for 5 days in an incubator at 28 °C in dark. After 5 days the fungal culture was inoculated on PDA (Fig. 12.2). Similarly, bacterial endophytes were revived on neem-based media and allowed them to grow for 2 days in an incubator at 37 °C in dark. After 2 days the bacterial culture was inoculated on LB broth and then inoculated on LA media. All the fungi and bacterial isolates were able to revived using this methodology. Also, the neem endophytes were capable of reviving back even after 4 years of storage in the cold (−80 °C) through neem media.

12.3 Bioassay of Fungal and Bacterial Endophytes Against Fungal Pathogen (*Magnaporthe Oryzae*)

The isolated endophytic fungi and bacteria were screened against the rice blast pathogen *M.oryzae*. *M. oryzae* is one of the potent plant pathogen causing the blast disease on rice plant and also in other cereal crops. This fungus causes about 50% of the rice yield loss. Therefore, it is very important to know the control agents which can reduce the pathogenicity so in this study we selected *M. oryzae* for bioassay against neem fungal endophytes. A loopful of bacteria was streaked on LB agar plate and incubated at 37 °C for 1 day whereas the fungal culture were inoculated at the four corners on PDA plate and

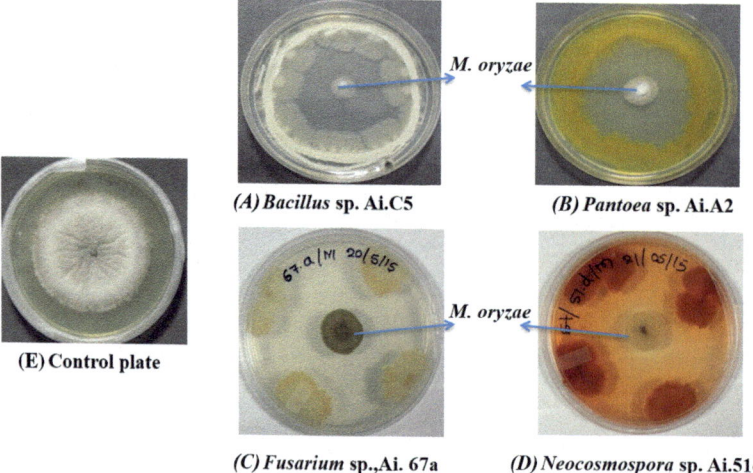

Fig. 12.3 Antifungal bioassay of neem endophytic bacteria and fungi against rice blast fungus, *Magnaporthe oryzae* grown on PAD media (Control). A loopful of bacteria was streaked at the periphery of the LB agar plate and fungal culture (*M. oryzae*) was inoculated on PDA plate

Table 12.3 16S rRNA and ITS sequence of endophytes showing antimicrobial activity against *Magnaporthe oryzae*

Bacteria/fungi	Isolate ID	Antifungal effect against *M. oryzae*	Closest NCBI match	Identity (%)	NCBI accession number
Fungus	49.d	+	*Xylaria apiculate*	99	KY031972
Fungus	51.d	+++	*Neocosmospora ramosa*	100	KY031973
Fungus	61.a	+	*Fusarium equiseti*	99	KY031974
Fungus	67.A	+++	*Fusarium sp*	99	KY031975
Fungus	F.10.A	+	*Aspergillus niger*	99	KY031976
Fungus	F.12.B	+	*Phoma sp.*	98	KY031977
Fungus	F.19.A	+	*Phaeoacremonium rubrigenum*	99	KY031978
Fungus	F.78.A	+	*Fusarium oxysporum*	99	KY031979
Fungus	F.78.B	+	*Fusarium oxysporum*	99	KY031980
Fungus	L.3d	+	*Alternaria sp.*	99	KY031981
Bacterium	C5	+++	*Bacillus sps*	99	KY029070
Bacterium	A5	+++	*Pantoea agglomerans*	98	KY029071
Bacterium	A2	+	*Pantoea sps*	98	KY029072

incubated at 28 °C for 7 days. After 7 days, the *M. oryzae* was inoculated in the center of these plates having endophytes. A control plate was kept wherein only *M. oryzae* was inoculated on PDA plates (Fig. 12.3). The control and bioassay plates were incubated at 28 °C for fungi and 37 °C for bacteria for 7 days. The growth inhibition of pathogenic fungi *M. oryzae* by bacteria and fungi was estimated by measuring the zone of inhibition (in cm). The experiment was performed in triplicates A total of 376 fungal and 80 bacterial endophytes were screened for the antimicrobial activity against the rice blast fungal pathogen, *M. oryzae*. There were 10 fungal and 3 bacterial strains suppressed the growth of *M. oryzae* (Fig. 12.3; Table 12.3).

12.4 DNA Fingerprinting of Fungal and Bacterial Endophytes

Endophytes showing maximum inhibition against rice blast fungus *Magnaporthe oryzae* were selected for phylogenetic analysis using 16S rRNA gene for bacteria and Internal transcribed sequence (18S rRNA ITS) sequencing for fungi.

12.4.1 DNA Isolation from Fungal and Bacterial Endophytes

Endophytic fungi were grown on PD broth for 4 days, then mycelium was harvested by filtering through muslin cloth. DNA was extracted from fungal endophyte using Gen Elute Fungal genomic DNA kit protocol (Sigma-Aldrich; NP-7006D). The DNA was extracted from the endophytic bacteria using Gen Elute Bacterial genomic DNA kit (W4502, Sigma-Aldrich). The 24 h-old bacterial suspension was centrifuged to obtain pellet which was further resuspended in 750 μL of suspension buffer (Chromas Biotech, Bangalore) along with 5 μL (20 mg/ml) of RNaseA (Qiagen Cat No. 19101) and incubated at 65 °C for 15 min. Then 1 ml of lysis buffer was added and incubated at 65 °C for 15 min followed by centrifugation at 13,000 rpm for 2 min. Then supernatant was transferred into two fresh tubes containing 900 μL of isopropanol and mixed gently. DNA pelleted by centrifugation at 13,000 rpm for 15 min and then the pellet was washed with 1 ml of 70% ethanol twice and air-dried at 37 °C for 10 min and dissolved in 50 μL of TE buffer.

Bacterial DNA was extracted from the endophytic bacteria using Gen Elute Fungal genomic DNA kit protocol (Sigma-Aldrich; NP-7006D). Then 24 h-old bacterial suspension was centrifuged to obtain pellet which was further resuspended in 750 µL of suspension buffer (Chromas Biotech, Bangalore) along with 5µL of RNaseA (Invitrogen) and incubated at 65 °C for 15 min. 1 ml of lysis buffer was added and incubated at 65 °C for another 15 min followed by centrifugation at 13,000 rpm for 2 min. The supernatant was transferred into two fresh tubes containing 900 µL of isopropanol and mixed gently. DNA was pelleted by centrifuging at 13,000 rpm for 15 min and the pellet was washed with 1 ml of 70% ethanol twice and dried at 37 °C for 10 min and dissolved in 50 µL of TE buffer.

12.4.2 PCR Amplification and Sequencing and Data Analysis

PCR amplification: For fungal identification, Internal transcribed sequence (ITS) was amplified using primers; ITS2-F (5'TCCTCCGCTT ATTGATATGC) and ITS5-R (5'GGAAGTAA AAGTCGTAACAAGG) (White et al. 1990), whereas 16S rRNA gene was amplified using universal primers; 16SV1-F (5'AGAGTTT-GATCMTGGGCTCAG and 16SV9-F(5'GGTT ACCTTGTACGACTT) for bacterial identification (Lane 1991). The PCR reaction was followed as per the protocol of Lane and coworker (1991) for 16S rRNA gene amplification and White and coworker (1990) for ITS gene amplification. The PCR reaction includes 1X Dream *Taq* Buffer, dNTPs (0.2 mM), primers (2 nM), template DNA (10 ng) and 1.25U of *Taq* polymerase (Fermentas) in a reaction volume of 25 µl. The PCR program includes denaturation at 95 °C for 3 min, followed by 35 cycles of denaturation at 95 °C for 30 s, annealing at 55 °C for 15 s for 16S rRNA gene amplification and annealing at 55 °C for 15 s for ITS amplification, followed by elongation at 72 °C for 60 s, with final extension at 72 °C for 5 min. The amplicons were analysed on 1% agarose gel and sent for sequencing.

12.4.3 Sequencing and Data Analysis

16S rRNA gene and ITS amplicons from these endophytes were sequenced at Sanger-sequencing facility, National Centre for Biological Sciences, Bangalore. The high-quality sequences were subjected to BLAST against the bacterial and fungal databases at NCBI.16S rRNA gene and ITS gene sequences of bacteria and fungi, respectively, were deposited in the GenBank nucleotide sequence database under accession no KY029070-72 and KY031972-81 (Table 12.3).

12.5 Analysis of Neem Metabolites from Fungal and Bacterial Supernatants and Pellets

The neem metabolite standards were purchased from different companies [azadirachtin (catalog number-PS-2075) (Sigma-Aldrich, Bangalore, India), nimbin (catalog number-N476280) (TRC, Canada), salanin (catalog number-19028-0173) (Chromadex, USA) and azadiradione (RD/NM/02) (Sami Labs, Bangalore, India)]. E/H-azadiradione was purified from the neem leaf extracts and gifted by Rangiah et al. (2016). Estrone-d4 was purchased from Steraloids Inc. (Newport, RI, USA). The purity of all analytes was >98%. High purity MS grade solvents (methanol, acetonitrile, and water) were obtained from Merck Millipore (Merck Millipore India Pvt. Ltd., Mumbai). Solid-phase extraction (reverse phase columns) was obtained from Phenomenex, Inc. (Hyderabad, India).

Two bacterial endophytes namely *Pantoea sp.* Ai.A2 and *Bacillus sp.* Ai.C5 and two fungal endophytes namely *Fusarium sp.* Ai.67A and *Neocosmospora ramose* Ai.51D were significantly inhibiting the growth of rice blast fungal in-vitro (Fig. 12.3) and were further selected for quantification of neem metabolite.

The metabolites were extracted from microbial pellet as well as supernatant as per the protocol of Rangiah et al. (2016).

12.5.1 Culturing Fungi

The fungal cultures were inoculated on neem plant-based media and incubated at 27 °C for 3 days. From the neem-based media, fungal cultures were transferred 50 mL of PDB (Potato Dextrose Broth) and incubated at 27 °C with shaking (200 rpm) on a rotary shaker for 4 days. Ten ml of incubated culture was filtered through the Miracloth to obtain the mycelial pellet and cell-free extract of fungus supernatant. The mycelia and supernatant (3 ml) were treated separately. The mycelial pellet was dried in speed vacuum and 10 mg of each sample was taken for further analysis. It was then resuspended in 50 µL of methanol and sonicated for 5 min to extract the metabolites.

12.5.2 Culturing Bacteria

The pure culture was prepared by inoculating the bacterial samples on neem plant-based media by incubating for 3 days at 37 °C. This was further transferred to 50 mL containing LB broth and incubated at 37 °C with shaking (200 rpm) on a rotary shaker till it reaches OD of 0.8 (Around 20 h). From this main culture, 1 mL was further sub-cultured in triplicate and allowed to grow in 50 mL of LB broth till it reaches the OD of 0.8 for 1 day. Then 10 mL of liquid culture from triplicate samples were centrifuged at 12,000 rpm for 10 min and the supernatant and pellet were separated. The pellet was dried in speed vacuum and 5 mg of each sample was taken for further analysis. This was resuspended in 50 µL of methanol and sonicated for 5 min to extract the metabolites. Then 3 mL of the supernatant was used to extract neem metabolites as mentioned below.

12.5.3 Extraction of Neem Metabolite

The 3 ml of supernatant from each bacteria and fungi was purified through reverse-phase solid phase extraction (RP-SPE) cartridges. The RPSPE cartridges were pre-conditioned with methanol (1 mL) followed by water (1 mL). Both fungal and bacterial supernatants (3 mL each) were loaded by adding 1 mL at a time and allowed to bind to the cartridge. To extract the neem metabolites, 50 µL of pellet extract was loaded on top of 1 mL of water in the cartridge after activation with 1 mL of methanol and allowed to bind. After loading, the SPE(solid phase extraction) cartridges were washed once with 1 mL of water and the neem metabolites were eluted from the SPE cartridges by using 1 mL of acetonitrile:methanol (80:20). It was then dried under speed vacuum and reconstituted with 50 µL of methanol.

12.5.4 Quantification of Neem Metabolite

The neem metabolites were quantified as per previously published UHPLC-MS/SRM method (Rangiah et al. 2016). About 10 µL of the above-prepared sample was spiked into 35 µL of methanol along with internal standard (5 µL estrone-d4 from100 µg/mL). The analysis was done by injecting 10 µL of the sample into the UHPLC-MS/SRM system (LC-Agilent 1290 infinity series, MS-Thermo Fishers TSQ vantage). The LC-MS conditions were same as previously published procedure (Rangiah et al. 2016). Since the methanol was used for the metabolites extraction, the calibration curves were prepared by spiking all five neem metabolites standards in 50 µL of methanol. The standards (10 µL) and the internal standard (5 µL from 100 µg/mL of estrone-d4) were spiked in 35 µL of methanol. The standards were passed through the RPSPE cartridge in the same way as mentioned above and final reconstitution was done in 50 µL of

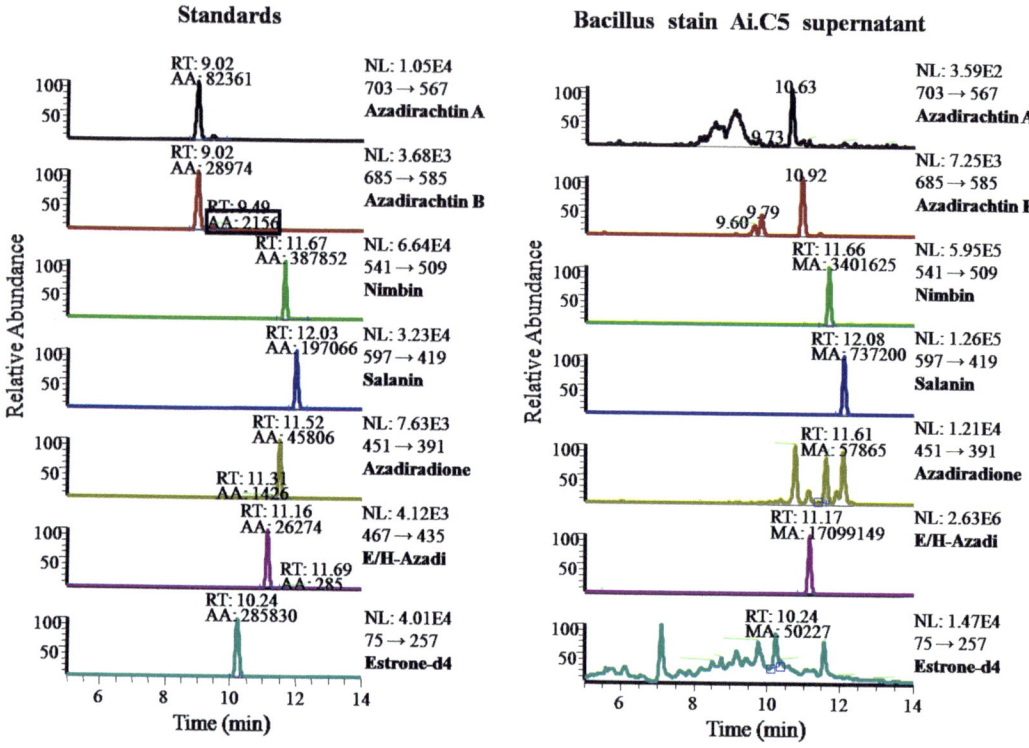

Fig. 12.4 The UHPLC-MS/SRM chromatogram of neem metabolites from *Bacillus sp.* Ai.C5 and standards

methanol. All five metabolites were analyzed under seven different concentrations. The amount on column was used to construct the standard curves are as follows: azadirachtin (78 pg, to 5 ng), nimbin (6.2 pg to 0.4 ng), salanin, azadiradione, and epoxy/hydroxyazadiradione (31.2 pg to 2 ng) as previously published protocol (Rangiah et al. 2016). The analysis was done by injecting 10 µL into the UHPLC/MS system. The area ratio (Analyte/ISTD) was plotted versus the concentration of analyte on the column.

Pantoea sp. Ai.A2 and *Bacillus sp.* Ai.C5 and two fungal endophytes namely *Fusarium sp.* Ai.67A and *Neocosmospora ramose* Ai.51D have inhibited Magnaporthe isolate. Then these bacteria and fungi were further used for quantification of neem metabolites (E/H-Azadi, Azadiradione, Nimbin, Salanin, and Azadirachtin) using UHPLC-MS/SRM method. The detailed method for metabolites identification and quantification was previously published (Rangiah et al. 2016). E/H-Azadi was the most abundant in these endophytes as compared to other metabolites, whereas Azadiradione and Azadirachtin were completely absent in both bacterial and fungal endophytes. *Bacillus sp.* Ai.C5 (Fig. 12.4), *Fusarium sp.* Ai.67A (Fig. 12.5) and *Pantoea sp* Ai.A2 (Fig. 12.6) produced 3 neem metabolites including E/H-azadi, nimbin, and salanin whereas *Neocosmospora ramose* Ai.51D (Fig. 12.7) produced only E/H-azadi. The metabolites were detected only in the supernatant but not in the pellet, thereby indicating that these metabolites are secreted out of the cells.

Interestingly, two of the bacterial endophytes, *Pantoea sp* Ai.A2, *Bacillus sp.* Ai.C5 and fungal endophyte, *Fusarium sp.* Ai.67A was showing production of 3 neem metabolites (E/H-azadi, nimbin, and salanin) at in vitro conditions, whereas fungal endophyte, *N. ramose* Ai.51D shown to produce only E/H-Azadi (Figs. 12.4, 12.5, 12.6 and 12.7). These metabolites secreted to culture medium and none found in pellet of

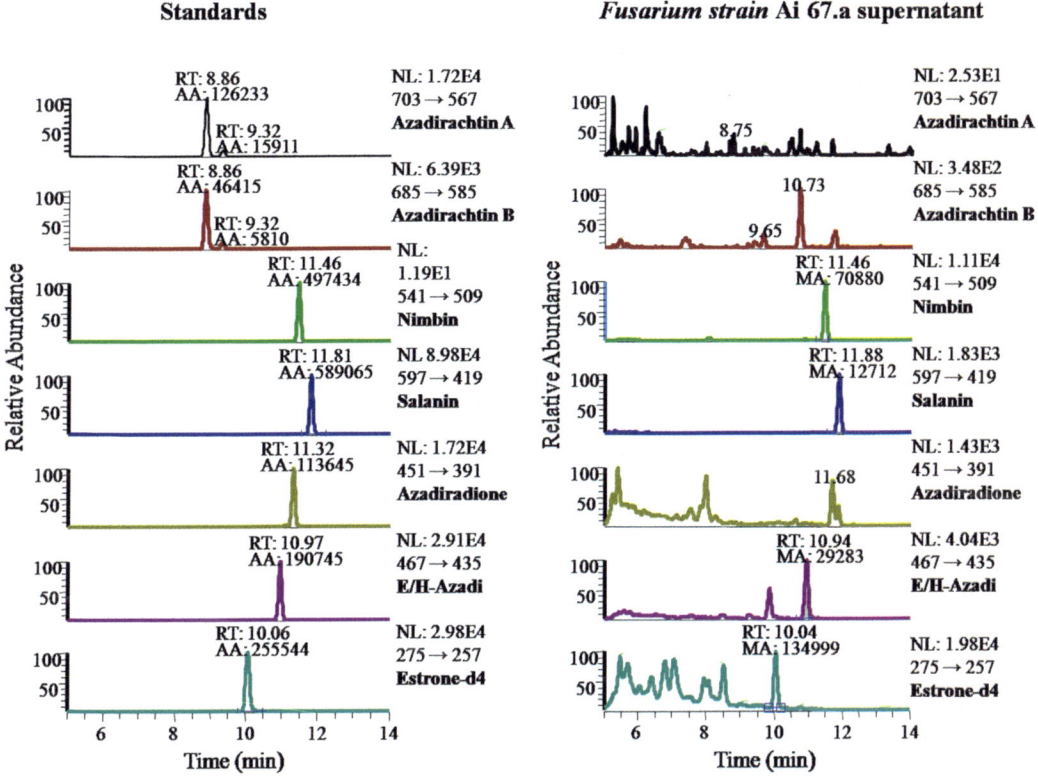

Fig. 12.5 The UHPLC-MS/SRM chromatogram of neem metabolites from *Fusarium sp.*, Ai.67A and standards

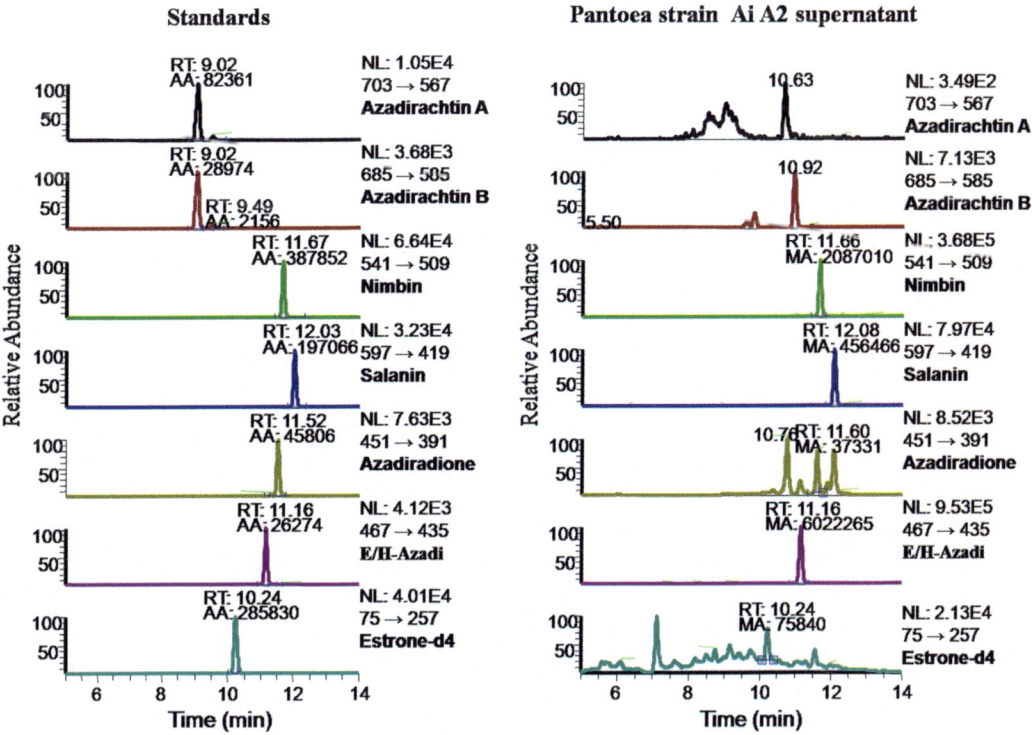

Fig. 12.6 The UHPLC-MS/SRM chromatogram of neem metabolites from *Pantoea sp.* Ai.A2 and standards

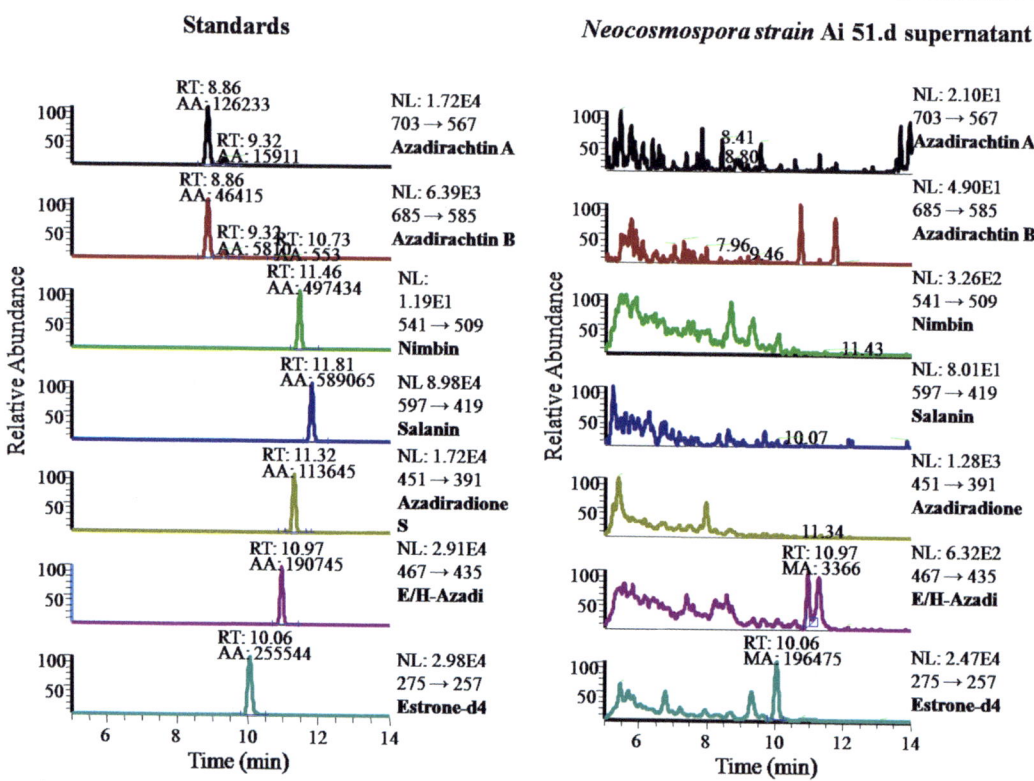

Fig. 12.7 The UHPLC-MS/SRM chromatogram of neem metabolites *Neocosmospora ramose* Ai.51.d(D) and standards

endophytes (Figs. 12.4, 12.5, 12.6, 12.7 and Table 12.4). This indicated that these metabolites are secreted out of the microbial cell, which make easier to devise a method for large-scale commercial production of neem tree metabolites at in vitro conditions. The recent study has also shown that neem endophytic fungus, *Eupenecillium parvum* produces the azadirachtin in an in vitro condition, which is similar to host plant (Verma et al. 2011). We did not obtain azadirachtin from the tested endophytes, however, we need to screen more fungal and bacterial strains from our collection to understand the metabolites diversity (Fig. 12.7).

We hypothesized that the metabolites encoding genes are present in both neem tree and also associated endophytes. Our hypothesis is supported by the previous findings of Taxol pathway, where Taxol synthesized by yew tree and its associated endophytic fungi, *Guignardia mangiferae* HAA11 (Tan and Luo 2011; Xiong et al. 2013). Taxol ($C_{47}H_{51}NO_{14}$) is widely used as anticancer molecule against breast cancer, lung cancer, and ovarian cancer. This created an alternative approach to obtain biomolecules through low-cost microbial fermentation process throughout the year (Stierle et al. 1993). The genes involved in metabolites pathway must have undergone evolutionary genetic recombination or exchange between endophytes to host plant and vice versa (Tan and Luo 2011). Endophytes from our study can be further analyzed for characterization of genes that are involved in metabolites synthesis pathway. This is useful in synthetic biology for large-scale production of plant secondary metabolites. Endophytes isolated from this study are maintained in their original status and can be used to

Table 12.4 Analysis of neem metabolites from bacterial and fungal endophytes

Neem endophyte	Source	Epoxy/hydroxyazadiradione (pg/ml)	Nimbin (pg/ml)	Salanin (pg/ml)	Azadiradione (pg/ml)	Azadirachtin (pg/ml)
Pantoea sp. Ai. A2	Supernatant	8,185,970	26,060	20,950	NF	NF
	Pellet	NF	NF	NF	NF	NF
Bacillus sp. Ai-C5	Supernatant	30,845,670	56,450	43,990	NF	NF
	Pellet	NF	NF	NF	NF	NF
Fusarium sp. Ai.67A	Supernatant	180.5	31.063	20.673	NF	NF
	Pellet	NF	NF	NF	NF	NF
N. ramose Ai.51D	Supernatant	16.9	NF	NF	NF	NF
	Pellet	NF	NF	NF	NF	NF

Note NF—Not found

12.6 Conclusions

A systematic study was carried out by isolating fungal and bacterial endophytes from different explants including leaf, bark, root, cortex, petiole, seed and fruit from 50-year-old mature neem tree. Around 376 fungal and 80 bacterial endophytes were randomly isolated and revived on neem-based media. These fungal and bacterial isolates were assayed for their antifungal activity against rice blast fungus *Magnaporthe oryzae* and further identified by 18S-rRNA ITS and 16S rRNA sequencing. Subsequently, these isolates were also profiled for presence of the bioactive compounds. Interestingly, these isolates were also producing the secondary metabolites in vitro that are reported to be produced by neem plant.

12.7 Future Work

Establishment of microbial repositories from the neem tree is an important step towards tapping their potential for human welfare, including drug discovery and sustainable agriculture. It remains to be investigated whether these organisms produce any novel secondary metabolites. Thus, produce large quantity of bio-pesticides, bio-fungicides, bio-bactericides anticancer molecules/drugs, bio-viricides, and antibiotics. these organisms are being explored for their secondary metabolites. In contrast, several other endophytes can be active against plant pathogenic fungi other than *M. oryzae*. These organisms need to be studied in detail and maybe exploited for disease management for important agricultural crops.

Acknowledgements We would like to thank Mahesh H B, Indira B.S, Shirke M, Yenagi V and Janani S for their help in endophytes isolation and DNA fingerprinting.

References

Brahmachari G (2004) Neem-an omnipotent plant: a retrospection. ChemBioChem 5(4):408–421

Debbab A, Aly HA, Edrada-Ebel RA, Müller WE et al (2009) Bioactive secondary metabolites from the endophytic fungus *Chaetomium* sp. isolated from Salvia officinalis growing in Morocco. Biotechnol Agron Soc Environ 13(2):229–234

Eevers N, Gielen M, Sánchez-López A, Jaspers S, White JC, Vangronsveld J, Weyens N (2015) Optimization of isolation and cultivation of bacterial endophytes through addition of plant extract to nutrient media. Microb Biotechnol 8(4):707–715

Lane DJ (1991) 16S/23S rRNA sequencing. Nucleic acid techniques in bacterial systematics, pp 15–175

Mahesh B, Tejasvi MV, Nalini MS, Prakash HS, Kini KR, Subbiah V, Shetty HS (2005) Endophytic mycoflora of inner bark of *Azadirachta indica* A Juss. Curr Sci 88(2):218–219

Rajagopal R, Suryanarayanan TS (2000) Isolation of endophytic fungi from leaves of neem (*Azadirachta indica*). Curr Sci 78:1375–1378

Rangiah K, Varalaxmi BA, Gowda M (2016) UHPLC-MS/SRM method for quantification of neem metabolites from leaf extracts of Meliaceae family plants. Anal Methods 8(9):2020–2031

Schulz B, Boyle C, Draeger S, Römmert AK, Krohn K (2002) Endophytic fungi: a source of novel biologically active secondary metabolites. Mycol Res 106(9):996–1004

Stierle A, Strobel G, Stierle D (1993) Taxol and taxane production by Taxomyces andreanae, an endophytic fungus of Pacific yew. Science 260(5105):214–216

Strobel GA (2003) Endophytes as sources of bioactive products. Microb Infect 5(6):535–544

Tan QG, Luo XD (2011) Meliaceous limonoids: chemistry and biological activities. Chem Rev 111(11):7437–7522

Verma VC, Kharwar RN (2006) Efficacy of neem leaf extract against its own fungal endophyte Curvularia lunata. J Agr Tech 2:329–335

Verma VC, Gond SK, Kumar A, Kharwar RN et al (2007) The endophytic mycoflora of bark, leaf, and stem tissues of *Azadirachtaindica* A. Juss (Neem) from Varanasi (India). Microb Ecol 54:119–125

Verma VC, Gond SK, Kumar A, Kharwar et al (2011) Endophytic fungal flora from roots and fruits of an Indian neem plant *Azadirachtaindica* A. Juss., and impact of culture media on their isolation. Ind J Microbiol 51(4):469–476

White TJ, Bruns T, Lee SJWT, Taylor JW (1990) Amplification and direct sequencing of fungal ribosomal RNA genes for phylogenetics. PCR protocols: a guide to methods and applications, vol 18, pp 315–322

Xiong ZQ, Yang YY, Zhao N, Wang Y (2013) Diversity of endophytic fungi and screening of fungal paclitaxel producer from Anglojap yew, Taxus x media. BMC Microbiol 13(1):71

Printed by Printforce, the Netherlands